TEXTS & DOCUMENTS

A Series of the Getty Center Publication Programs

The TEXTS & DOCUMENTS series offers to the student of art, architecture, and aesthetics neglected, forgotten, or unavailable writings in English translation.

Edited according to modern standards of scholarship and framed by critical introductions and commentaries, these volumes gradually mine the past centuries for studies that retain their significance in our understanding of art and of the issues surrounding its production, reception, and interpretation.

Eminent scholars guide the Getty Center for the History of Art and the Humanities in the selection and publication of TEXTS & DOCUMENTS. Each volume acquaints readers with the broader cultural conditions at the genesis of the text and equips them with the needed apparatus for its study. Over time the series will greatly expand our horizon and deepen our understanding of critical thinking on art.

BUILDING IN FRANCE
BUILDING IN IRON
BUILDING IN FERRO-CONCRETE

Published by the Getty Center for the History of Art and the Humanities

BUILDING IN FRANCE BUILDING IN IRON BUILDING IN FERRO-CONCRETE

SIGFRIED GIEDION

TEXTS & DOCUMENTS

INTRODUCTION BY SOKRATIS GEORGIADIS

TRANSLATION BY J. DUNCAN BERRY

THE GETTY CENTER PUBLICATION PROGRAMS
Julia Bloomfield, Thomas F. Reese, Salvatore Settis, *Editors*
Kurt W. Forster, *Consultative Editor,* TEXTS & DOCUMENTS

TEXTS & DOCUMENTS
Architecture
Harry F. Mallgrave, Editor

Building in France, Building in Iron,
Building in Ferroconcrete

Stanislaus von Moos, Editorial Consultant
Verena Clay, Translation Consultant
Lynne Kostman, Managing Editor
Benedicte Gilman, Manuscript Editor

Published by
The Getty Center for the History of Art and the Humanities,
Santa Monica, CA 90401-1455

Printed in Canada

01 00 99 98 97 96 95 7 6 5 4 3 2 1

Cover: Bruce Mau Design (based on fig. 58, p. 145)

Translated from
Sigfried Giedion, *Bauen in Frankreich, Bauen in Eisen, Bauen in Eisenbeton.*
1st ed. Leipzig: Klinkhardt & Biermann, 1928.

Library of Congress Cataloging-in-Publication Data is to
be found on the last printed page of this book.

Contents

Preface ix

Sokratis Georgiadis — Introduction 1

SIGFRIED GIEDION
BUILDING IN FRANCE, BUILDING IN IRON, BUILDING IN FERROCONCRETE 79

INTRODUCTION 85–100

CONSTRUCTION 87
INDUSTRY 88
ARCHITECTURE? 90
CONSTRUCTOR AND ARCHITECT 94
CONSTRUCTORS 97
THE NINETEENTH CENTURY 99
THE NINETEENTH CENTURY IN FRANCE 100

IRON 101–149

1800–1850 FIRST ATTEMPTS 103
FIRST FORMATION 110
1850–1890 EXPERIMENTAL ARCHITECTURE 120
EXHIBITIONS 120
EXHIBITIONS OF WORLD COMMERCE 122
FURTHER DEVELOPMENT 143

FERROCONCRETE 150–204

NATIONAL CONSTANTS 152
1900–1920 A. G. PERRET 154
TONY GARNIER 161
1920–1927 LE CORBUSIER AND THE YOUNGER GENERATION 167
THE CURRENT STATUS OF FERROCONCRETE 200

Appendix 207

Illustration Credits 225

Sigfried Giedion: A Biographical Sketch 226

Index 228

Preface

Every effort has been made in this translation to follow the style of the German text as closely as possible. This strategy was in fact made necessary by the care that Giedion himself lavished on his text: his deliberate use of images, capitalization, boldface, and emphasis on individual words or phrases as a polemic in itself, that is, as an effort to structure the text in an almost hierarchical manner, so that the reader might visually discern the essential components of each page. The veracity in this regard also extends to Giedion's particular syntax or style — for instance, when he sometimes intentionally omits verbs — or when certain key words specifically evolve from his polemic. Thus *Konstrukteur* has been translated as "constructor" rather than the more common "engineer"; *Eisenbeton* has been translated (following Giedion himself) as "ferroconcrete" rather than "reinforced concrete." It is hoped that in this way some of the tone of Giedion's original thought has been captured.

There are no italics in Giedion's original text or notes. His bibliographic references and his footnotes have been presented largely as he gave them, with only minimal editing to make them intelligible to the modern reader. It has not been possible to verify all quotations, so they are reproduced as Giedion presented them. Obvious typographical errors relating to matters of grammar, spelling, geography, names, and dates have, however, been corrected in the translation. French words and quotations have also been edited to conform to standard usage.

In various respects this text is very much a joint venture. I would like to thank Harry Mallgrave for his significant participation in the making of this book, Stanislaus von Moos for important comments and suggestions, Benedicte Gilman for her editing and revisions to the translation, and Lynne Kostman for her guidance of the book through its editorial phase. I would also like to express my appreciation of the work of J. Duncan Berry and David Britt, who translated from the German and French, respectively. Also special thanks to Verena Clay and to Maria Georgiadou, both of whom contributed greatly to the accuracy and quality of the final translation.

For a complete listing of Giedion's publications, the reader is referred to the bibliography compiled by Stanislaus von Moos and published in *Hommage à Giedion: Profile seiner Persönlichkeit* (Basel: Birkhäuser, 1971), 187–98.

— S. Georgiadis

Introduction

Sokratis Georgiadis

There is broad agreement among historians of modern architecture that Sigfried Giedion's *Bauen in Frankreich, Bauen in Eisen, Bauen in Eisenbeton* (1928) may without reservation be called a classic of modern architectural literature (fig. 1). Reyner Banham presents this book as a kind of authorized history of Modernism: "the Modern Movement's view of its own history."[1] On the basis of this book and in connection with Giedion's work as secretary of the Congrès Internationaux d'Architecture Moderne (CIAM), Spiro Kostof characterizes him as "the official historian of the Modern Movement."[2] And Stanislaus von Moos maintains that with *Bauen in Frankreich* Giedion has "drawn the picture of a developmental continuity" that "time and again has inspired the history of modern architecture."[3]

Nevertheless, one can hardly describe Giedion's intention as primarily historiographic. With this book, born as it was from a series of articles written for the journal *Cicerone* and still bearing "the odium of a manifesto" (von Moos), the author wanted to be, in his own words, "the conveyor of a viewpoint." He wanted to sketch a vision of a modern architecture that — in accordance with the demands of the avant-garde — broke decisively with the traditional values of the discipline. Only later did he place this vision within a historical context. He attempted to do so by means of the hypothesis that the great iron constructions of the nineteenth century and the ferroconcrete architecture produced by the pioneers of Modernism belong aesthetically under one and the same roof.

Intellectually, *Bauen in Frankreich* is the product of Giedion's direct contact with the leading personalities of Modernism, "with the artists of his own time." This contact began during the summer of 1923 with the large Bauhaus exhibition in Weimar (his first meeting with Walter Gropius) and continued during the *Esprit Nouveau* exhibition of 1925 in Paris (his first contact with Le Corbusier). It resulted in Giedion's numerous articles on contemporary architecture, which appeared in newspapers and professional journals in Germany and Switzerland. With the publication of *Bauen in Frankreich* and with the assumption of his role as secretary general of CIAM (which he helped found in 1928), Giedion the historian renounced the neutrality of an outside observer of events; he willingly shed his supposed

Fig. 1. Sigfried Giedion. Courtesy Archiv S. Giedion, Institut für Geschichte und Theorie der Architektur, ETH-Hönggerberg, Zurich.

innocence and became, both programmatically and definitively, an integral part of the Modern Movement. This, in any case, is how Giedion himself related the sequence of events on various occasions. Yet things began quite differently.[4]

Giedion was a Swiss citizen born in Prague on 14 April 1888. After completing his secondary schooling, he studied mechanical engineering in Vienna at the behest of his parents. But engineering, with a view to taking over the family's textile concern, was not to his liking, and he decided instead to study art history. In 1915 he commenced his studies at the University of Zurich and eventually transferred to the University of Munich, where Heinrich Wölfflin, one of the great art historians of the day, taught. Giedion did his doctoral work under Wölfflin and in 1922 published his dissertation, *Spätbarocker und romantischer Klassizismus* (Late Baroque and Romantic classicism).[5] Concurrent with his studies he wrote poetry, prose, and—inspired by the Aktivismus literary movement—a play called *Arbeit* (Work), which was staged in 1917 in Vienna, Leipzig, and Basel, and in Berlin by Max Reinhardt. In the turbulent years following World War I Giedion became politically active and supported the Munich *Räterepublik*. In Munich he also met the art historian Carola Welcker, who was to become his wife.

Notwithstanding his excellent preparation for it, an orderly academic career interested Giedion about as much as the management of a textile mill. Instead he cast himself in the role of a comrade-in-arms in and—wherever possible—fellow creator of an admittedly still-vague but all the more passionately desired cultural renewal. It was precisely with this aim that he turned toward modern architecture, which received its first, unshakable foundation with *Bauen in Frankreich*.

In the decade following its publication, Giedion was active as a freelance writer on art. He contributed regularly to the journals *Cicerone* and *Cahiers d'art;* he wrote a small book on the modern housing form for a Zurich publisher (*Befreites Wohnen* [Liberated dwelling], 1929); and for Editions Crès he prepared the first short monograph on Gropius (*Walter Gropius,* Paris, 1931). He also worked intensively on a projected multivolume history of modern civilization, *Die Entstehung des heutigen Menschen* (The origin of modern man), which remained unfinished. From his home in the Zurich Doldertal, which served both as the CIAM headquarters and as a meeting place for Europe's avant-garde in art and architecture, Giedion launched an extraordinarily broad series of activities. He organized exhibitions, gave numerous lectures, and codirected a firm that had as its goal the dissemination of modern furnishings.[6] His interests ranged from the origins of the new vision in modern painting to efforts to introduce a quiet toilet. As an architectural client he initiated construction of two prototypical multifamily residences in Zurich, designed by Alfred and Emil Roth and Marcel Breuer.[7]

In the 1920s Giedion also tackled the issue of architectural education. In the early 1930s he even sought an academic post at Zurich's Eidgenössische Technische Hochschule (Swiss Federal Institute of Technology, ETH). His breakthrough, however, did not come until a few years later when, through the auspices of Gropius, he was appointed the Charles Eliot Norton Professor of Poetry at Harvard University for the academic year 1938–1939.[8] From his series of lectures titled "The Life of Architecture" came the book *Space, Time, and Architecture,*[9] which has gone through numerous editions and been translated into several languages. One of the century's most successful architectural books, it gained a considerable international reputation for its author. Giedion integrated large parts of *Bauen in Frankreich* into this work, which Gropius described in a letter to its author of 18 May 1941 as "really the best statement about our movement."

In *Space, Time, and Architecture* Giedion also broadened his conception of Modernism. He discussed the turn-of-the-century's moral revolt against eclecticism and the revolution of modern painting (which was the first art to develop a new spatial conception) as further important factors in defining the new architecture — factors that reached beyond a simply technical-constructional interest. In precisely this architecture, the cultural role of which extended beyond the narrow boundaries of the discipline he had pondered, Giedion believed he could discern the first signs of a synthesis of the objective, practical, and expressive aspects of culture and of a repeal of modern man's disastrous "schism between thinking and feeling," for which he blamed the nineteenth century.

This dilution of the subject matter of *Bauen in Frankreich* can be traced not least to the fact that in *Space, Time, and Architecture* Giedion made an important new discovery: America. In quick "close-ups" he sketched the architecture of the New World, from the anonymous buildings of the first settlers to the so-called Chicago school and Frank Lloyd Wright. America now became for Giedion an aspect of the modern adventure that could not be circumvented and at the same time a "primary influence in the maturation of his historical vision."[10] It is thus not surprising that America would become the central theme of his next book, *Mechanization Takes Command.*[11]

Whether in the Patent Office in Washington or in the slaughterhouses of Cincinnati, he steadfastly sought out the documents of an anonymous development that impressed itself upon the countenance of modern life. Meanwhile, Europe degenerated into a negative image of an emotionally unresolved onslaught of the modern. The vestiges of *Bauen in Frankreich* had now completely disappeared, and the optimistic perspective that had defined the tone of this publication in the 1920s similarly disappeared. Giedion questioned the omnipotence of a rationalism that was an end in itself and registered his skepticism of the naive faith in progress. In his view, the command of mechanization went hand in hand with the increasing loss of humanity, which ultimately became apparent in the unparalleled catastrophe of World War II.[12] Its restoration became the primary objective of his work for the next twenty years, until his death on 9 April 1968.

During these last years Giedion's interests followed two paths. He continued to be an activist in the Modern Movement, directing CIAM until its dissolution in 1956. He developed new ideas for a "more humane" Modernism and concerned himself with such questions as the "new monumentality," the "new regionalism," the increasing significance of aesthetic values in architecture, and the possibilities of architectural expression.[13] At the same time, he continued his historical research, the results of which directly influenced his teaching, which now alternated between the ETH-Zurich and the Harvard University Graduate School of Design. The origins of art and the beginnings of architecture now became the themes of his studies.[14] Constancy and change became the main aspects of his focus. It was his avowed goal to make visible the immutable elements of the human condition embodied in art and architecture leading up to Modernism.[15] The avant-garde pathos of the 1920s, which resonated throughout *Bauen in Frankreich,* was thereby silenced. Perhaps this reflected the fate of Modernism as a whole.

I.
"An Almost Aesthetic Pleasure": The Architectural Appropriation of the New Materials and *Bauen in Frankreich*

The Crystal Palace, designed by Joseph Paxton, was erected for the London World Exhibition of 1851. The fascination and shock provoked by this building were due not least to the fact that it conveyed the impression that architecture could free itself from its material constraints and even to a large extent dissolve itself as a body. No one summed up the relevant feelings better than the architect Gottfried Semper, who called it a "glass-covered vacuum."[16] This sensation was made technically possible by using a material that seemed quite unexpectedly to shatter the age-old tangible certainty of architecture. This material was iron.

Some eight decades separate the erection of the Crystal Palace and the appearance of *Bauen in Frankreich*. This is, by and large, also the period during which the debate over the architectural use of iron took place. Whereas the debate was in part very intense and indeed passionate, it also had quieter stretches and more sober turns. Nevertheless, much was at stake: nothing less than a new style of architecture, the new form from which one hoped to tailor an architectural dress for the modern age. Semper's ideas about iron formed one of the two most important reference points during the first phase of this debate. The other pole was defined by the theories of Carl Gottlieb Wilhelm Bötticher, the author of the epoch-making study *Die Tektonik der Hellenen* (The tectonics of the Hellenes).[17]

Iron: The Expectation of the New Art-Form

Bötticher's decisive plea for the extensive architectural use of iron and this material's promise of creating a new style go back to his lecture "Das Prinzip der Hellenischen und Germanischen Bauweise hinsichtlich der Übertragung in die Bauweise unserer Tage" (The principle of the Hellenic and Germanic styles of building with regard to their application to today's way of building). It was an address given at the first festival commemorating Karl Friedrich Schinkel, organized in 1846 by the Berliner Architekten-Verein (Berlin architectural association).[18]

In his monumental work *Die Tektonik der Hellenen,* Bötticher had concerned himself with comparing "Hellenic and Germanic styles of building." As he had noted in that work, he felt obliged "to dispel the suspicion that a constrained, one-sided preference for the forms of Hellenic tectonics has guided the principle of my work and robbed me of an eye and free awareness of the value of medieval tectonics."[19] Bötticher's contribution can thus be placed within the context of the ongoing debate on style, which had excited the temperaments of architects and architectural thinkers since the 1820s.[20] He sought to reconcile the differences by proposing a stylistic synthesis. In this, Schinkel's work provided him with a prototype.

But in his Schinkel address of 1846 Bötticher developed a completely different position, the foundation of which was a radical historicizing of the architecture of antiquity and Gothic times. On this ground he questioned the viability of either style for future development. Bötticher's starting point was the same one he had used in *Tektonik der Hellenen:* architecture had no analogs in the world of perception, no models in nature. This was the reason for its "working independence." He reproached classicists and Neogothicists alike for the fact that they had remained stuck on the "outer shell" of architecture. The "working principle" of architecture — from which, so to speak, the different physiognomies of the two styles had arisen — rested precisely on their respective "static principles" and "material relations." For this reason, the two competing parties had failed to see that the two ways of building had been merely evolutionary stages of a history that would certainly not end there.

> Both [styles] define only two evolutionary stages, which have to run their course and complete their prescribed circle before a third style can see the light of day. This in no way invalidates either of the styles; rather, only by basing itself on the results of both can a third and higher stage of development come to the fore — a third style, whose creation is reserved by historic necessity for a later time but whose foundation our own age has, in fact, already started laying.[21]

Classic and Gothic architecture are no longer seen, as in the *Tektonik der Hellenen,* as dialectical opposites, whose poles are on the verge of synthesis. The historical perspective opens up, and this opening is bound to the certainty of an expected epochal change, upon the threshold of which Bötticher believed his generation had already arrived.

Bötticher was convinced that iron represented the dawn of a new, higher historical level. He believed this material presaged a new, third way of building and consequently also a new architecture. Naturally, he made his point of view from a constructional perspective. Just as he had asserted in *Tektonik der Hellenen,* he attached decisive importance to roof construction and regarded it as the most essential aspect of an architectural style. The system of monolithic stone beams in antiquity and the arched vaulting of the Middle Ages, he said, had completely exhausted the constructional and architectural possibilities of stone. A material was now needed that could facilitate construction and span great openings, one that could therefore guarantee a reliable roofing system and thus permit every imaginable floor plan and spatial configuration — a material that consequently might bring with it a "new realm of art-forms." This material was none other than iron.

Bötticher was certainly not the first to discuss the new material from a theoretical perspective. The year before Bötticher's Schinkel address, Eduard Metzger contributed to the still-lively debate on style, and iron construction constituted an important part of his argument. Metzger described the view of his contemporaries thus: "*Iron Constructions:* I can well imagine these words terrify the plastically minded architect!"[22] But he was of the opinion that the architect must face the challenge presented by iron. He himself did so by seeking to link the use of iron with his own penchant for the Neo-gothic. "The network of the Gothic arch and all of its spanning members is closely related to the one that develops in accordance with the nature of iron."[23] In Metzger's eyes, the main problem was the practical joining of iron to stone construction, for "it almost appears as if iron spanning members and the more massive architectural bodies are distinct incompatibilities."[24] Although he offered no solutions, he spoke confidently concerning the developmental possibilities of a new pointed-arch style in which iron would play a critical role.

The significance of Bötticher's contribution is already evident. He was the first to deduce a wholly new tectonic principle from iron, one fundamentally different from any that had gone before. At the same time, he endeavored to base this principle on a historical-theoretical foundation. He did not, however, succeed in satisfactorily answering the aesthetic question. Although he held out the prospect that iron construction's "core-form" (*Kernform*) would create a new sphere of "art-forms" (*Kunstformen*), he avoided outlining even roughly the character of these art-forms, which he had done in *Tektonik der Hellenen* for stone buildings of antiquity. There the art-form was the symbolic representation of the tectonic form, or the core-form. Richard Streiter, who in 1896 wrote a comprehensive critique of Bötticher's *Tektonik,*[25] pointed out this omission and correctly noted that the various attempts to translate Bötticher's theory into practice had not yet yielded "a fundamentally new way of designing, evolving from the characteristics of material and technique."[26] Instead, Bötticher's students had simply transposed the ornamental motifs of the Greek temple to the iron components.

"Invisible Architecture"

Bötticher's remarks did not meet with unconditional agreement. Even Heinrich Leibnitz, who in a short book of 1849 attempted to popularize the *Tektonik der Hellenen* and apply its system to all of architectural history, distanced himself from Bötticher's position as soon as it touched upon the architectural applicability of iron. At the end of his book, where he examined contemporary events, Leibnitz noted that "iron is not the kind of material that can call forth a new artistic epoch in architecture, for it transgresses measure, contradicts static feeling, and destroys mass."[27] The question of the new style was also drawn, like a red thread, through the entire theoretical work of Gottfried Semper. His most brilliant statement on this subject was given in his lecture of 4 March 1869 in the Zurich City Hall, in which he discussed the conditions of the origin and change of architectural styles.[28] His polemical sting was directed against two contemporary theoretical conceptions of style. On the one hand, he criticized the historical Darwinists who would apply the laws of natural selection, heredity, and adaptation to architecture. On the other hand, he opposed those who "start from the erroneous assumption that the question of style is chiefly a constructional question." It is clear already from this criticism that Semper's position is incompatible with the idea that a constructional material alone has the capacity to create a style — as Bötticher's Schinkel address implied.[29] The same is true for Bötticher's prophesy that the use of iron could lead to a third, modern way of building, following the Greek and Gothic styles.

Semper's negative attitude toward the extensive architectural use of iron was also based on another important factor. He had discussed the material as early as 1849 in an article reviewing the Jardin d'hiver in Paris,[30] and there he altogether rejected the open display of it, at least in monumental buildings. As a constructional material iron should "by its nature be used as slender rods and sometimes as cables." Due to the small surface displayed in these forms, the more perfect the construction, the more the iron visually disappears. It is in this sense that Semper spoke of it as an "invisible material." Barely fifteen years later he repeated this assessment in his principal work, *Der Stil in den technischen und tektonischen Künsten; oder, Praktische Ästhetik* (Style in the technical and tectonic arts; or, Practical aesthetics). We have in metal-rod construction, he noted, "a poor soil for art! There can be no question of a monumental style with cast-iron rods; its very ideal is *invisible architecture*! For the thinner metal is spun, the more perfect is its manner."[31] Clearly Semper could not tolerate a dematerialized architecture. For him, matter carried the symbolic form of the architecture and thus constituted a necessary precondition for the pursuit of its social function.

This is why Semper completely dismissed the idea that iron could engender a new monumental architecture and warned that "the dangerous idea that iron construction, applied to monumental building, could give us a new architectural style has already led many talented, yet artistically estranged, architects astray."[32] It is also quite certain that here Semper had in mind Henri Labrouste, whose Bibliothèque Sainte-Geneviève in Paris he had also explicitly criticized in his review of 1849.[33]

Whether in the form of rods or cables, iron's unfitness for monumental purposes did not mean, however, that it was unsuited for other tasks. Semper was fully prepared to acknowledge the possibility, for instance, of using the material for decorative purposes in latticework. In addition, he advocated the use of iron in sheet form. By that he meant "sheet beams" (*Blechbalken*), which must always remain invisible but could play an important role in bridging large spans; sheet metal used as a wall dressing or as a material in doors; and finally hollow, cast-iron forms, such as tubular columns.

Moreover, iron could be used in buildings of a nonmonumental nature, in Semper's words, "buildings of a decidedly practical purpose." The iron roof trusses of railroad stations gave him a "satisfactory impression." In *Der Stil* he spoke about them in an almost enthusiastic way but distinguished all the more rigorously such building tasks from monumental architecture: "Let us permit and praise the visible, plain iron roof trusses of the railroad engineer in terminals and other such things as symbols of their provisional nature. Spare us the wonders of decorated iron libraries, festival halls, and the like!"[34] The consequence for the architect is clear: he should not yield to the temptations of the material.

Iron: The Engineer's Material

The Gotha architect Ludwig Bohnstedt[35] reflected on the premise that architects and engineers had completely different responsibilities, which in many respects could never be reconciled. This premise also influenced, as it were, his ideas on the architectural use of iron. With clear reference to Bötticher's Schinkel address of 1846, Bohnstedt thus noted in 1867 in his essay "Über die Bedeutung des Eisens für die Baukunst" (On the significance of iron for architecture): "The question of the creation of a new style that owes its origin to metal may be seen as already settled; a new style stemming from the use of iron cannot be expected."[36]

Yet, at the same time he recognized iron's applicability to the constructions of engineers, and he even spoke of the great progress that could be made when iron was used in appropriate works. The strength of the material and its ease of handling were its greatest advantages. Speaking about architecture, he maintained that iron was completely inappropriate for carrying the art-form because "it had to appear undisguised." Like Bötticher, Bohnstedt was convinced of the role construction played in determining a style. He believed, however, that only stone construction (that is, the quintessential architectural material) was receptive to a formal treatment, and he could not imagine the possibility of forming a style without taking this into account. "Our traditional laws of style," he wrote, "are rooted precisely in our experiences with a solid material — with stone — and have been made to harmonize with it; those laws determine the fulfillment of all demands, which up to now *only stone* has been able to satisfy."[37] Consequently, only stone was acceptable as a construction material; stone was the "autocratic ruler," as Bohnstedt characteristically expressed it. The reason for iron's unsuitability for architecture was its lack of corporeality, the impossibility of employing it "in all dimensions." Its volume could only expand and approach that of stone when it was shaped as a hollow body, but that would contradict its materially appropriate use.[38]

Preliminaries Toward an Incorporeal Architectural Aesthetic

The aesthetic development of iron construction was the topic of a lecture delivered in 1870 to the Berliner Architekten-Verein by Richard Lucae, architect of the Frankfurt Opera House.[39] It was not the first time he addressed this issue. A year earlier he had written an article for Andreas Romberg's *Zeitschrift für praktische Baukunst,* in which he entered what was at the time uncharted theoretical territory. The subject was the architectural significance of space, which the author attempted to decipher by examining various architectural tasks. Among these were some that were directly related to the use of iron.[40] At the center of Lucae's reflections was the effect of space on human feelings. He saw form (understood as the relationship of spatial dimensions to one another), light, measure, and color as critical factors for spatial perception as well as decisive artistic moments. Each of these factors played a particular role as a component of architectural space. Thus, form was responsible for the aesthetic effect of space; light for its character; color for the specific mood produced; and, finally, measure was "the spatial relation of our body to our spirit as perceived by our consciousness." Lucae specifically emphasized that the question of style could be ignored within the framework of his theory: "Style only influences the effect of space to a very small degree."[41] He gave different explanations for this, but failed to mention the most important: namely, that his theory of space removed from style (in the sense of conventional historical styles) the basis on which it had traditionally been able to work — the architectural body. And it was precisely this that led him to iron construction. Thus his extreme reticence is quite surprising in his discussion of the spatial effects of those building tasks for which iron is the material of choice. Among these are the railroad station halls, in which, according to Lucae, one "had thought one could do without art."

> If a meaningful thought of beauty were added to the grand constructional ideas of their roof form, then our eye might find rest and enjoyment among the confusing web of iron bars and cables crossing each other in every direction. Then we would, figuratively speaking, not notice the individual instances of this mathematical formula translated into iron and simply organize the result into an easily surveyed sum, into a system appearing as beautiful form. For the purely mathematical construction is not a finished artistic achievement, but only a skeleton like the one in the human body.[42]

Lucae, on the other hand, was utterly fascinated by the Crystal Palace, which in the meantime had been rebuilt in Sydenham. He characterized the building as a "piece of sculptured atmosphere," as a "magically poetic form of light," and forgot, apparently out of sheer enthusiasm, about the demand for a "meaningful thought of beauty." He elaborated upon his remarks in his lecture on the aesthetics of iron construction. Lucae referred to the issue as "one of the most important questions of architectural development," in order to underscore the great difficulties one encountered in surmounting it. These difficulties were without exception related to the diminished bulk of iron construction compared to stone construction. Lucae again used the analogy of the human body: "*A purely mathematical construction is as much an unfinished artistic result as is the human body with its muscles and ligaments exposed, or even as its skeleton is capable of being a living creation of nature;* thus I maintain that the beauty of a building system is partially due to the fact that there is a surplus of mass beyond the material necessary for support."[43]

Lucae thought that architects were facing a dilemma. Through its "limited corporeality" iron had rid itself, "as it were, of the material by which we can display beauty." Without an increase in iron's dimension, it lacked the aesthetic sense conditioned by stone. Increasing its bulk, on the other hand, robbed it of its characteristic property, "namely, that it remains delicate and yet still evokes the impression of strength." Yet Lucae did not exclude the possibility that "a new art-form" could arise "out of the nature of iron";

rather, he shifted its emergence into the future. His own generation was denied the push toward the new art-form as well as an unbiased approach to the question of iron construction: "A succeeding generation will grow up with iron construction, just as we have grown up with stone construction; they will in many cases have the fully undisturbed sense of beauty that still today leaves us unfulfilled, for we feel that the tradition of beauty so dear to us is under attack."[44]

From this basic assessment of contemporary conditions Lucae deduced the promise of a new art-form: his was a technical age rather than a religious one, an age dominated by exact science and its representatives, the engineers. As a result, the hub of the new architectural aesthetics could be neither the palace nor the church, but solely utility buildings — that is, bridges and transportation buildings, the vessels of modern life.

He sought to sketch the elements of the new sense of beauty and support it with his previously developed theory of space. He was thoroughly convinced that the creation of space was not the result of constructional considerations. The "totality" of space, the "spatial image," originated exclusively in the human imagination. Although construction remained architecture's sine qua non, it was inconceivable without a fundamental idea of beauty. Lucae was of the same opinion as Bötticher (without mentioning him by name) that the roof was of decisive tectonic importance. At the same time he wanted to see Bötticher's concept of *Tektonik* expanded. It was one of the principles of tectonics that the art-form had to explicate symbolically the constructional conditions, yet this was far from being its only task: "But tectonics glimpses its higher task in the grand, general thoughts that must be expressed with each small, individual symbol — the words of these thoughts, so to speak!"[45]

In the case of the roof, for example, this general symbol was "suspension." (It is interesting to note that the "suspension" of Greek tectonics loses its original meaning as a structurally passive architectural element and is here transformed into a symbol.) To this end and in order to bolster its true significance, Lucae thought it possible that the actual construction should not be visible: "As paradoxical as it might at first appear, one can only show the roof if one more or less conceals it."[46]

Lucae's theory broke up the unity of body and space. It had still existed for Bötticher, who wrote in the preface to his *Tektonik der Hellenen* that "the working hand of the tecton [builder] forms every member according to a corporeal schema as it... fulfills the creation of space."[47] He meant that the "space-creating organism" was none other than the corporeal members (*membra*) of that organism. In drawing upon this, he maintained in his Schinkel address that the "essence of architecture resided in expanding" ideas into a "corporeal-spatial entity."

What is significant about Lucae's understanding of the roof as a symbol of "suspension" is that the "spatial image" in this instance no longer needs the body for its support; at least Lucae in this instance refrains from referring to the need for corporeality. This conception comes very close to an aesthetic of the noncorporeal as it was developed by later Modernists.

These ideas indicate the general limits within which Lucae saw the role of the architect. The architect was not allowed to operate exclusively based on his "constructional feeling"; he needed to collaborate with the representatives of the exact sciences. However, his specific role was indispensable. Lucae stood for the "separation of disciplines," even though he was also convinced that the architect and the engineer were working toward a common goal.

Iron: Emaciated Stone

At the third convention of the Verband der deutschen Architekten- und Ingenieur-Vereine (League of German architects' and engineers' associations), in 1878, Semper's biographer, Constantin Lipsius, delivered a lecture entitled "Über die ästhetische Behandlung des Eisen im Hochbau" (On the aesthetic treatment of iron in architecture).[48] In it he attempted to articulate a modern position on this issue, a position that would be open to the most recent developments. The premises upon which the science of engineering produced its works were presented as follows:

> Among the technical sciences, the science of engineering, a pure child of our time, has forcefully stepped to the fore. Standing upon thoroughly modern, thoroughly real grounds, it aims at the fulfillment of purpose and—delegating aesthetic questions to other enterprises—strives to express them with a relentless consistency and the most naked, inexorable truth. The less material expended or the attainment of maximum results with minimal dimension, the greater the triumph! And since purely technical purposiveness does not require transfiguration through beauty, since its visual appearance is only the expression of the function it performs, that is, function that has become form, such purely purposive artifacts convey in their construction also the explanation of their existence, their necessity; in that way they are often aesthetically satisfying and convincing to a certain extent.[49]

But precisely only to a certain extent, for in order to evoke aesthetic pleasure, the satisfaction of emotional requirements should be added to that of intellectual ones. Lipsius believed that truth and beauty were mutually dependent, yet a truth oriented merely to purpose such as that expressed in the works of engineers is not sufficient for a work of art. The real truth must be supplemented by an ideal truth. And this is precisely the domain of the architect, whose task it is to endow the ideal with artistic form.

Lipsius warned all those who were "promising to themselves the blossoming of a new, unique art from the use of iron" against exaggerated expectations. Although elements of a new worldview loomed on the horizon, he added (in following Semper) that its particular architectural dressing did not yet exist. Yet, during the process of its formation one could not deny a certain importance to the emerging technical possibilities corresponding to the use of the new materials—naturally presuming that its aesthetic implementation be successful.

Returning to the original question of the aesthetic treatment of iron, Lipsius then explained that he did not mean its use as a constructional aid but its appearance as an independent organism. He first of all drew a sharp distinction between wrought and cast iron. To the former he attributed no significance for monumental architecture beyond its use in the manufacture of light latticework. The slight volume of wrought iron lagged behind its actual capacity for carrying a load. With cast iron it was different; when formed as a hollow body, it assumed a "corporeality in this manner of presentation similar to that of stone, the monumental building material par excellence." It was suitable for both columns and trusses and also for the construction of roofs. The important thing was to make the cast iron itself recognizable and visible. Only in this way could "the possibility of a building's solidity" be visually understood and its "factual reality" explained. The same was true for the connections between the individual constructional members and for the connections between supporting and supported parts. In the discussion of roofs, whose importance he (like Bötticher) stressed, Lipsius praised the achievements of Labrouste (referring to the Bibliothèque Nationale in Paris) and Félix Duban (referring to the roofing of the inner court of the Ecole des Beaux-Arts), and he praised one of the iron designs presented by Eugène-Emmanuel Viollet-le-Duc in his *Entretiens.*

Lipsius's concept of the "aesthetic treatment of iron" remained fixed to the tectonics of stone construction. Iron appeared as a substitute for stone;

iron construction was nothing more and nothing less than a stone building with emaciated contours.

Although the display of iron construction for Lipsius was an essential precondition for its aesthetic effect, its appearance belonged merely to the preliminary stages of the architectural work of art, which he viewed as the aesthetic embodiment of an architectural idea beyond construction. The aesthetic realm remained the exclusive domain of the architect. Thus the goals of the architect and the engineer were in his view widely divergent. At best, their mutual contributions could be compared as creative and inventive activities.

From Invisibility to Monumentality

At the tenth annual meeting of the Verband der deutschen Architekten- und Ingenieur-Vereine, in 1892, the architect Hubert Stier delivered a lecture entitled "Rückblick auf die Entwicklung der deutschen Architektur in den letzten 50 Jahren" (Retrospective of German architectural development in the last fifty years).[50] Referring to the question of the architectural use of iron, he spoke of two phases during this period. In Stier's words, the first phase began "with the use of iron mainly as cast iron for decorative purposes instead of wrought iron, whose treatment had been forgotten; there was at that time hardly a glimmer of understanding of iron's importance for constructional purposes."[51] The entire first phase, which according to Stier lasted until 1872, was admittedly characterized by an increasing use of iron, "but primarily only as a constructional aid under the sheathing and veiling of another material. No one had the courage to display it openly. One wrote about its aesthetic treatment, but no one attempted such treatment."[52]

The debate about the use of cast and wrought iron was in fact settled at the latest in the early 1880s, and indeed in favor of wrought iron. A commentator in the *Deutsche Bauzeitung* noted that the hope was long gone "that through cast iron one could achieve either new stylistic forms or transformations of certain elements of inherited stylistic forms according to time and place, such as, for instance, the Moorish."[53] Architects now favored wrought iron alone, the "so-called genuine" material.

But wrought iron still posed a problem, and possibly an even greater one than cast iron. One was still uncertain as to how to treat it aesthetically. Its limited corporeality or linearity offended many people, as did the "confusing number of small, crisscrossing structural parts — flat bars, angle irons, round bars, and so on."[54] There was a demand for visual repose, which one hoped to achieve, for instance, by using supports covered by compact decorated walls instead of latticeworks and frames. A further nagging problem was the treatment of the points at which iron joined the stone construction; this problem appeared especially in cases where wrought iron by itself was used to cover vast spaces.[55]

The Cologne architect Georg Heuser also attempted to solve the troubling problem posed by Semper and others since him—architecture's loss of corporeality through the use of iron. At the end of the 1880s and in the early 1890s, Heuser launched an extended journalistic investigation into the aesthetic question of iron construction.[56] He developed his position from his reading of Bötticher's *Tektonik der Hellenen,* Semper's *Der Stil,* and Darwin's theory of evolution. He was particularly interested in "compartmented forms" (*Gefachformen*), that is, iron construction consisting of the "flange and web" (*Gurt und Steg*). These elements furnished him with the name of his theoretical construction: the "compartmented style" (*Gefachstil*). Heuser proposed a three-step plan by which the dematerialized lattice-and-rod construction visually transformed itself into a mass-form, which only then made possible a monumental architecture:

> Depending on the shape of the web, the expression of the compartmented style changes as follows:
>
> 1. If the web consists of the lattice-and-rod system that characterizes the "iron style," then we have an *invisible architecture*...
> 2. If the web is a thin but generally *enclosing wall,* then we have *a bodiless, materially weak, yet very visible architecture,* even in iron...
> 3. If the web is massive in whatever material, then we have a *corporeal, monumental architecture,* in which the principle of "flange and web" can be appreciated as an artistic idea, both for its *stability* and for its basic decorative richness.[57]

In this way, Heuser traversed the path from line to surface to mass, thereby intending to eliminate Semper's reservations about iron construction. The only thing that seemed to remain was a convincing answer to the question of decoration. And precisely on this issue, Heuser attempted by means of Darwin to confirm Bötticher's art-form. Heuser championed the naturalistic ornament, which was applied to the technical form—the latter understood as "organ projection" or the imitation of nature—and which was thus analogous to the static function of each constructional member. Here the decoration was subordinate to the constructional web.

Despite his very ambitious theoretical claims, in the final analysis Heuser's position remained rather conventional. The aesthetic closeness of iron and stone, which he explicitly presented as a goal, in the end amounted to a perversion of the new construction.

Heuser's "compartmented style" found approval and support from the architect of the Imperial Post Office in Berlin (*kaiserlicher Postbaurat*), Robert Neumann.[58] In his remarks on the future architectural style, iron played an extremely important role. The well-known themes were repeated in his enumeration of the advantages and disadvantages of iron. Iron construction was lauded for its strength, durability, and ability to span wide spaces. Its lack of corporeality, however, its light and airy character (which did not allow the affixing of decoration), limited the possibilities of its use. Although the author praised Heuser's compartmented style, he registered doubts about iron's ability to accomplish monumental tasks. Neumann's argument is at least paradoxical: he thought that iron used structurally in roofs accomplished too much, far more than needed for monumental purposes. Hence, he concluded: "Iron in monumental buildings will probably always be used mainly in roofs, thus falling far short of taking full advantage of its structural potential; for that reason, one should not overrate the style-forming influence of iron."[59]

Yet Neumann also saw great potential in iron when it was used in conjunction with "hardened mortar materials," that is, with cement. Otherwise, one could really only use exposed iron in architecture for the roofing of interior spaces. "On the exterior of buildings," notes Neumann, "iron will not be very conspicuous, and for this reason it will have little influence on exterior architecture. Here stone, and above all the very efficient ashlar, still retains its old right."[60]

"Scaffold Styles" — "Mass Styles"

Despite the contradictions and difficulties in Heuser's theory, iron construction was still the point of departure for his compartmented style. This was not so for all the defenders of architectural corporeality. On occasion, the fixation on stone assumed lofty tones. This building material became, in fact, a synonym for architecture, and efforts were made to link it inseparably with the historical destiny of the discipline.

At the end of the 1880s the architecture professor Adolf Göller in Stuttgart confessed his belief in stone. In his book *Die Entstehung der archi-*

tektonischen Stilformen (The origin of architectural styles),[61] he laboriously laid down fourteen "new" formative principles by which a style allegedly undergoes historical change. As soon as he started discussing iron in the conclusion, however, he fell into astonishingly conventional categories of thought. He essentially repeated the reproach of "bodilessness" and expanded it by claiming that no decorative forms could be developed from wrought iron and therefore an aesthetic treatment was hardly possible. He also rejected the connection between iron and stone as a basis for the development of a new style:

> Indeed, a new style for our time is, despite some hopes, not to be expected from this union; for one thing, in most building types wrought iron will never accomplish very much for wall construction; for another, with its slightness iron can never offer sufficient *form*. Therefore, stone must, as before, be the sole vehicle for such a style.[62]

The defense of stone against iron led many writers into theoretical sophistry, to claims that could not stand up to even elementary historical or factual review. One such example was given by the editor of the *Deutsche Bauzeitung,* K. E. O. Fritsch.

Fritsch delivered the keynote address at the ninth annual meeting of the Verband der deutschen Architekten- und Ingenieur-Vereine in Hamburg in August 1890.[63] He was interested in stylistic changes in contemporary architecture; however, he placed the theme within the historical perspective of architectural development since the Enlightenment. According to Fritsch, the first phase of this period was characterized by the "idealism" of early historicism. Around 1850 this was replaced by an increasingly realistic attitude, due in part to the significant impact of Gottfried Semper. The "realism" of the second phase consisted of different styles. Fritsch listed them—Hellenistic, Renaissance, northern Renaissance, Baroque, Empire, Gothic, Romanesque—and did not conceal his preference for the Neoromanesque; yet he came to the conclusion that none of the currently used styles could claim absolute validity. In closing, he expressed the wish for a new style, which he was inclined to see as a welding together of all historical styles. In this connection, he also spoke of the "iron style." Fritsch emphatically rejected the idea that this could become the future style, as had been said everywhere since the Paris World Exhibition of 1889: "I must confess to being a stubborn heretic on this point. For such a belief seems to me to be not only an extreme overestimation but above all a complete misjudgment of the significance of iron for architecture."[64]

Fritsch's attack was directed at none other than Bötticher and his Schinkel address of 1846. Fritsch rejected Bötticher's thesis that the roofing system was the historical motor of architectural change — a thesis from which the latter had concluded that iron would be of decisive importance for a future style. Fritsch held that the wall rather than the roof was the decisive element. From this perspective, the history of architecture recognized two main groups of styles. He called them, respectively, "scaffold styles," which derived from the craftsmanship of carpentry, and "mass styles," which derived from masonry. In ancient Egypt and Mesopotamia, in Rome and in the early Middle Ages, and also during the Renaissance and the Baroque, mass styles were practiced without exception. Ancient Greek and the Gothic architecture, on the other hand, were scaffold styles. In the period immediately after the Enlightenment, the scaffold style became the foundation of architectural activity. Fritsch saw this simply as an enfeeblement and gave the following explanation:

> The reason, however, that the stylistic experiments that started more than a hundred years ago were oriented primarily toward the two scaffold styles — Greek and Gothic art — is simply that these styles, as opposed to the mass styles, had a *canon*. However, the canon yields to a certain extent because it is necessary to let the architectural scaffold appear; the mass style, on the other hand, allows considerably more freedom to the individual sentiment of the artist. Weaker talents are thus always more comfortable employing a scaffold style, and it was also inevitable than an age that demanded, above all, firm principles for artistic activity eagerly reached for them.[65]

It is surprising, in fact, that the realist Fritsch praised artistic individuality and clearly renounced the idea of architectural normativity. The upshot of his analysis was that the artistic freedom he so desired could be attained only through the material that permitted the effect of mass, in short, stone. Iron merely permitted a scaffold construction. In order to be able to argue against iron, Fritsch, as expected, quoted Semper: iron offered "a poor soil for art," and its use resulted in an "invisible architecture." An "iron style" was nevertheless possible, in Fritsch's view, but only in the sense of a marginal event within the architectural panorama.

The Beauty of the Line

The division of responsibilities between architect and engineer — and consequently the question that had been dramatically addressed by Bohnstedt in the late 1860s, only to be taken up again later by Lucae and Lipsius — raised further and no less serious concerns for architects of the 1890s. One can assume that this problem was one of the reasons many architects maintained an air of reserve toward the new material. The defense of architecture's domain in the face of the ingress of iron — the engineer's material — was in some respects also the defense of the professional interests of the architect against the increasing power of the engineer. Conversely, in view of iron's inexorable triumphal march, one began with the conviction that the position of the architect could only be secured if the discussion of iron's use in architecture were uncoupled from simple utilitarian considerations and diverted to a course in which aesthetic arguments prevailed.

Stier had the following to say concerning the period 1870 to 1890:

> Iron, too, has not only assisted us in creating ever freer spaces in dimensions that had previously been thought impossible: we have also attended to its artistic development. I must say, however, that precisely in those areas where the two professional lines of art and engineering must work together, such a collective effort has been somewhat lacking in Germany.[66]

He hoped for a future cooperation of the two disciplines and regretfully observed from the Paris World Exhibition of 1889 that the French had "somewhat anticipated" this issue. The importance that Stier attributed precisely to this problem appeared in the concluding lines of his fifty-year retrospective, in which a subliminal sentiment of professional antagonism toward engineers is unmistakable:

> May our younger sister, the science of engineering, in short order have dashing success: we will happily and ungrudgingly grant her that, for we shall also reap great advantages from it. Yet the architect's patent of nobility has been written by thousands of years of human cultural history, and if in fifty years a speaker shall look back to today, he will confirm that we, too, have understood how to uphold that nobility![67]

A short while later, the same problem concerned the architect (and later coeditor of the *Deutsche Bauzeitung*) Albert Hofmann: "Is it possible that

Hercules the engineer sits at the distaff of the delicate beauty Omphale, that the liberal and cheerful art of architecture can get in touch with the somber austere seriousness of the science of engineering?"[68]

Thus he poses the question and seeks an answer in light of the global upheavals caused by the intellectual and material conditions of the time. The author's conclusion, based on an anatomy of the contemporary zeitgeist, is that under the new conditions, the advance of the engineer seemed not only plausible but an outright necessity. "Everything is in transition," wrote Hofmann, "and this ongoing transition is at the same time economic, scientific, political, and philosophical."[69] In his opinion the old idealistic worldview was in the process of disintegration, induced by the pressures of the "positivism of science," the "nihilism of intellectuals," and the "materialism of daily habits." Metaphysics had been replaced by "independence of thought, the scientific boldness of research, and industrial and commercial activity." Taking the art historian Robert Dohme's remarks in *Das englische Haus* under advisement, Hofmann thought that these changes influenced human experience.[70] Thus the expectation for new forms was not directed toward the architects, nor was the demand for a new architecture corresponding to these needs. "The engineer," Hofmann continued, "presses forward victoriously, and it is undeniable that the world is enthralled by his bold and beautiful works."[71]

Implicit in these statements is the question of the aesthetics of engineering construction, which, as Hofmann further argued, was none other than the aesthetics of iron construction. The engineer's works are inextricably bound to iron. Stone, according to Hofmann, is the domain of the architect; iron, by contrast, has "put the development of technology primarily in the hands of engineers." Can iron construction be beautiful? "Can iron bridges be beautiful?" asked Hofmann along with the mechanical engineer and cultural theorist Franz Reuleaux.[72] He was inclined to answer the question with a rigorous "Yes!" By this beauty, however, he did not mean "the free, picturesque or so-called artistic beauty, but rather a dry, mechanical beauty deriving from the structural framework." But which of the two beauties is the most beautiful? "The *architecte constructeur* was always the soul, the *architecte décorateur* the dress for this soul," wrote Hofmann, leaving no doubt that ultimately it was the soul that mattered.[73]

> Today, in many cases, the dress has become the main thing, the soul has been abducted by another.... The architect is in danger of sinking, the engineer

> ascends disproportionately. This sharp, calculating, rational, intellectual activity, supported by the steadily developing sense of beauty, forms the sublime moment in contrast to the free, artistic, emotional activity of the architect.[74]

The new feeling of beauty brought forth by the engineer was founded upon the "awareness of the beauty of the line." The beauty of the line is precisely what aesthetically separates engineering construction from the works of the architect. The beauty of the line was indeed not unknown to the architect, but it did not play the same critical role in his creations as it did in iron constructions. The two concepts of beauty were for this reason different but still not incompatible. But when they were combined, "it was certainly indisputable that the beauty of the line forms the highest concept of beauty."[75] With this, Hofmann made his aesthetic preference clear. However, his aim seems to have been the unification of architecture and engineering in a synthesis, and he saw this synthesis as the retreating architect's only chance of survival. He explicitly distanced himself from the position of Lipsius, who had postulated the divergence of the purposive and the beautiful in his lecture of 1878. Hofmann saw the harmony of their relationship in the much more comprehensive unity between art and exact science: "Until now, art has allowed itself to be led more by feelings, [but] why should it not make the intellectual progress of the exact sciences its own? Here opposites clash and are united, feeling and thinking cling to the same appearances and are reconciled."[76] This occurs because both serve the new culture. The same is true for architecture, whose creations must, on the one hand, satisfy the needs of the age and, on the other hand, bear its characteristic features.

> Necessity will create a *rapprochement* out of the *contrast* between the two; today's rivals will work together tomorrow. And a new art arises from the collaboration of taste and mathematics, of formal beauty and technological boldness — a picture of our culture, like Euphorion, the symbolic son of Faust and Helena, in whom is united the spirit of the old and the new age, the enthusiasm for beauty, and the sense of the utilitarian.[77]

The Immanent Beauty of the Work-Form

Despite the different views in the debate about iron, one thought seems to form the lowest common denominator for the various positions: it is agreed that in order for a building to be characterized as an aesthetic object, the architect must always do something special, beyond merely meeting functional needs or constructional efficiency. Even for Hofmann, who dismissed the sharp distinction between purposive form and beauty, the sense of beauty remains a necessary pendant to the rational activity of erecting a building. Left to his own devices, the engineer is incapable of producing beauty. Beauty belonged exclusively to the domain of the architect.

The idea of an immanent beauty of purposive form became increasingly important in the next phase of the argument and heralded a shift that characterized several contributions. There were, of course, important differences among this community of authors. There were those who began with a causally acting mutual dependence between purposiveness and beauty. To them, purposive form was aesthetic precisely because it was purposive. Others conceded that beauty had a certain autonomy. In order for purposive form to be beautiful, it was necessary to activate a specific sense of beauty in the early phases of the object's design: in this way one could summarize the relevant views. Even with these authors, however, the ability to create beautiful form ceased to be the privilege solely of architects. The engineer increasingly appeared on the stage with a claim to equal rights.

The engineer Gustav Lang addressed this theme very early, in an essay published in 1891.[78] He pointed to the approximate character of structural theories, which in themselves could supply no clues about possible failures or inexactitudes of calculations. This task fell to a "healthy sense of beauty," which at the same time served as a criterion for judging the correctness of the mathematical calculation. This was possible because correct theory and a healthy sense of beauty were mutually dependent. According to Lang, the constructions of engineers were by no means things in themselves "that could be neither beautiful nor ugly"; they were instead "man-made works that had to reflect those natural laws in which our whole feeling and thinking are so rooted that nothing can be held to be beautiful that appears purposiveless according to these laws."[79] And he asserted apodictically that

> a contradiction between the products of correct mathematical calculation and a healthy sense of beauty does, in fact, not exist. . . . If our mathematics leads to ugly forms, it means that not all the relevant influences have been correctly balanced in the setting-up of our formulae; the sense of beauty may thus remain for us a guiding star that prevents our mathematics from becoming one-sided.[80]

The postulate of an immanent aesthetics of purposive form was a symptom of the increasing approval for the wide-ranging use of iron in architecture. The reasons for this are certainly not to be sought or found in a successful conversion of iron construction into architecture, or rather, aesthetics. Iron construction found acceptance, above all, because it became regarded as a genuine expression of the modern zeitgeist. In a practical way, the aesthetic sense had to comply with this turn in cultural history. Cornelius Gurlitt found an accurate expression for this fact: "We have not artistically conquered the work-form of iron," he wrote, "rather, the work-form of iron has conquered us and forced us to see it as beautiful, for it is rational and the product of a creative idea."[81]

In 1890 Gurlitt settled his account with Bötticher: praising him as the last man of the Enlightenment but at the same time rejecting him as "a representative of a scientific art." "Hellenism is out of the question for the German nation," Gurlitt continued; he explained this by the extinction of the "rationalism of art." He then announced prophetically that "the days of individualism will dawn again!"[82] What Gurlitt did in 1890 with Bötticher, he repeated ten years later with Semper. He maintained that "Semper's entire theory of style" was demolished because we have ceased to comprehend the symbolic content of a work of art as the "application of stylistic forms." He attached to his criticism against Semper his disagreement with Otto Wagner's thesis that in architecture the art-form gradually arose out of the work-form. This criticism implied the view that the work-form by itself already possessed both aesthetic and symbolic relevance (arising from Gurlitt's criticism of Semper).

For Gurlitt, the new paradigm was Japanese art, whose forms "are simply developed from laws of utility and are yet undeniably artistic." Moreover, they can convey symbolic content. This is the basis on which Gurlitt, so to speak, integrated the forms "invented by engineers" into the realm of art. He vehemently opposed the view that "the work of the mathematically inclined engineer was inherently inartistic" and that consequently the architect has "to make forms intelligible through art." Straightforward design, practical planning, and simple and purposive execution suffice to characterize a building as beautiful. Even the general population understands this. Naturally this explanation brought Gurlitt into conflict with his earlier criticism of Bötticher and his alleged "scientific art." For the beauty of which Gurlitt himself spoke, as he conceded, was a pure product of mathematical calculation. Be that as it may, Gurlitt's contribution is typical of the new ideas that had now entered the debate. The magical words "utility," "*Sachlichkeit*" (architectural realism), and "purposiveness" — with all of their implications — now set the tone.

Hermann Muthesius, for example, emphasized in his famous book *Stilarchitektur und Baukunst* (Style-architecture and building-art) that modern architecture had to conform to the new economic and transportational conditions, to the new constructional principles, and to the new materials. He praised the architecture of iron and glass, the works of Paxton and Labrouste, the Eiffel Tower and Galerie des Machines of the Paris World Exhibition of 1889, and finally Alfred Messel's Wertheim Department Store (1898) in Berlin. But somewhat later in the text he explicitly deduced from these givens the signs of a new style, indeed, with a wording that fully endorsed Gurlitt's conceptualization:

> If we wish to seek a new style — the style of our time — its characteristic features are to be found much more in those modern creations that truly serve our newly established needs as for example in our railway terminals and exhibition buildings, in very large meeting halls, and further, in the general tectonic realm, in our large bridges, steamships, railway cars, bicycles, and the like. It is precisely here that we see embodied truly modern ideas and new principles of design that demand our attention. Here we notice a rigorous, one might say scientific objectivity [*Sachlichkeit*], an abstention from all superficial forms of decoration, a design strictly following the purpose that the work should serve. All things considered, who would deny the pleasing impression of the broad sweep of an iron bridge? Who is not pleased by today's elegant landau, trim warship, or light bicycle?[83]

Muthesius returned to the theme of iron construction in an essay of 1913, published in the yearbook of the Deutscher Werkbund.[84] His position was different from that of 1902 in that he now no longer traced the cause of aesthetic feeling (the pleasing impression) experienced through technical form to the fulfillment of practical purpose. He argued, like Hofmann in 1893, that a certain sense of form contributed to the technical form of iron construction and that it, essentially, helped to shape the appearance of iron. He was even more specific in linking this sense of form to an "aesthetic subconsciousness," which was active in every human form, such as the "production of instruments, building, and constructional activity." As he noted, "Even if the producer aimed only at purposive reason, we could claim that he has also been influenced by considerations about form, even if only through an unfelt aesthetic 'subconsciousness.'"[85]

Muthesius now gave an unequivocally negative answer to the question that Hofmann left unanswered in 1893, namely, whether purposiveness and beauty are mutually dependent. "It is a false idea that an engineer is entirely pleased when a building, an instrument, a machine that he creates satisfies a purpose; still more erroneous is the recently often-heard proposition that if it fulfills a purpose, it is also beautiful. Utility in and of itself has nothing to do with beauty."[86]

And precisely because they are independent of one another, it is necessary to weld them together. This synthesis does not take place when the architect beautifies the exterior of an engineer's work. The sense of form has to contribute at the outset to the transformation of the purposive idea that supports the object, as a kind of controlling device. The logical conclusion is that Muthesius did not make any distinction between the works of engineers and of architects, and in this he once again distanced himself from Hofmann's position of 1893.

Muthesius consequently rejected the notion that the works of engineers were ugly and that beauty was therefore the province solely of the architect. He distanced himself from Semper's criticism of iron construction. According to Muthesius, "massive" stone architecture, which contrasted with iron's "invisible architecture," was nothing more than an ideal formed by a habit and brought about by the fact that one had always built with materials that seemed massive. Yet, this should in no way be taken as an absolute ideal.

Incidentally, Hofmann had by 1907 reached a view very similar to that expressed in Muthesius's text of 1913. He spoke of a necessary "spiritualization of matter," an operation that already participated in the conception of the building and that transcended the simple consideration of purpose. Parallel to this he stressed that "we expect from engineering the enrichment of art through new forms and new designs, as external expressions of a new power inherent in the human intellect and its productions."[87]

An argument similar to that presented by Muthesius in his book of 1902 was put forward by Friedrich Naumann, the other spiritual father of the Deutscher Werkbund. He, too, fought against the belated artistic treatment of purposive form and believed that because of its purposiveness, such form was already beautiful. Concerning iron construction, he wrote: "Here art is not practiced alongside construction: no pasted-on decoration, no simple curlicues; here creation is for a purpose, and form is born like a child unplanned by its parents."[88] Naumann thereby indicated the aesthetic unconditionality of engineered form, and hence its actual detachment from any genealogical connection to the discipline of architecture. He then plastically described the emancipation of technical form from its incapacitation by historical architecture — an emancipation that heralded a new era of civilization. In another passage, Naumann simply referred to this era as the "iron age."

> Untold potential exists here. All the old concepts of space are shifting. Relations of support and load have changed. To let huge vaults rest virtually on points is something so new that the architect, as if ashamed of his youthful strength, still often finds false piers necessary. Out of a kind of timidness, iron buildings are still given stone vestibules. Yet the magic of the new art resides precisely in this quiet and yet so happy emergence from the forest of the past.[89]

In the first years of the new century Heinrich Pudor also published a series of essays on iron construction — mainly in the journal *Der Architekt* as well as in a separate brochure of 1904, entitled *Die neue Architektur.*[90] He causally deduced the aesthetic effect of a technical object (whether a machine or an iron construction) from the form's immediate accord with its practical purpose. The basis of aesthetics was the fulfillment of this purpose with the minimum means. For Pudor "means" signified above all the material necessary in the production of the object. He thus expected an enhancement of the aesthetic effect through a reduction of the material expended, understandably with a concomitant enhancement of the quality of the material, which he understood as durability and efficiency. He called this the "material style," noting that it "frees us from superfluous decoration and ornament and combines beautiful purposive form with beautiful material," and that it "constitutes the modern form of expression not only in the applied arts but also in commercial art and industry, and likewise in architecture and engineering."[91]

To support his thesis, Pudor used a biological analogy. He criticized every attempt to imitate stone architecture in iron construction and contrasted it with the ideal of an "airy," frame-and-skeletal, iron architecture. He spoke of the "architectural revolution" in progress, made possible by iron. It unfolded in two phases: "Here, too, the first period is characterized by the fact that one was ashamed of iron construction and tried to conceal it from view, whereas in the second period, whose beginning is marked by the Eiffel

Tower, its naked beauty is exposed to view."[92] In general, the Eiffel Tower was of paradigmatic importance to him. It signaled a historical epoch that was just beginning.

In the essay "Ästhetik des Eisens" (The aesthetics of iron), Wilhelm Freiherr von Tettau also occupied himself with the well-known problematics: "Should technically perfect expression be identical with artistically perfect expression?"[93] Not quite, was his answer, for this overlooked the actual distinction between science and art. In structural processes the "power of inner forces" was active, and this was precisely what should be brought out and articulated. This was the task of art. The inner tension carried to the surface appeared as an interplay of concealment and pronouncement, and was "a matter only of artistic tact." That which was statically incomprehensible could be made more apparent to the spectator through artistic expression. Only through this was aesthetic pleasure possible, for "I am convinced that aesthetic pleasure is increased with greater comprehension of the system."[94]

Contradiction

These views could not be established without being contested. Richard Streiter, for instance, expressed grave doubts. In his book *Architektonische Zeitfragen* (Contemporary architectural questions),[95] which was conceived as a critique of Otto Wagner's *Moderne Architektur,* he nevertheless took positions on all the important themes of the contemporary architectural debate. He summarized the discussion on iron construction since Bötticher—rejecting both the latter's prediction that a "new realm of art-forms" could be expected from the new constructional material and Wagner's statement "that wherever art has shaped this material completely new forms have in fact appeared. Thus it has provided one of the greatest impulses to the growth of the new style!"[96] Finally, Streiter also called to account Heuser's "compartmented style."

Not only was Streiter's own position characterized by the idea of the relative autonomy of artistic form from technical form, but he even gave the former unequivocal preference. Although he accepted the possibilities advanced by the new technical means—he described the extensive use of iron as a building material as "the greatest constructional achievement of the modern age"—he was also convinced "that the new technology will not generate a new style, but that a certain sense of form will artistically develop the new technology in this or that way."[97] Streiter had some doubt as to whether this was indeed possible with iron, that is, whether it lent itself to artistic treatment or at least allowed such treatment. His response to this question was similar to Semper's argument, but it also combined various other arguments critical of the architectural use of iron into a platform for its rejection. Understandably, the defense of the architectural body, the sine qua non of architecture's artistic effect, assumed center stage.

> Wherever iron structures of significant dimension stand visible and alone, their absolute resistance to artistic design becomes apparent. The hope that the future will bring the solution to something not yet achieved can certainly be described as illusory. For the possibility of developing an iron construction through a structural-symbolic shaping of the whole, and especially of the parts, into a no longer abstract-rigid system but into a concrete-living structural organism, one whose members by the expression of their corporeality allow us to feel immediately their cooperation in a "happy state of equilibrium" — this single possibility of the tectonic "idealization" of the work-form is precluded by the nature of the material and by the nature of its assembly. The fleshless thinness and stiff tediousness of structural parts; the system's adherence to structural calculation; the external uniformity of the elements that in general disallows recognizing the differences of their load (resistance to tension or compression); the confusing quantity of crisscrossing, almost incorporeal lines in large constructions, whose sense and purpose can be understood only by a technically trained reasoning but not by simple feeling: all this makes iron construction appear indifferent to us — even though, it must be admitted, there is often a certain aesthetic charm in the overall contour of such works (arched bridges, the Eiffel Tower) or in the effect of the enormous interiors they make possible. Yet even the grandest iron structure will not succeed in evoking an important, profound, and truly artistic mood.[98]

Opposition to iron construction was voiced not just by those who wanted to bring the institution of architecture (with its basic aesthetic assumptions and normative positions) safely into the new century. Opposition also came from those movements forming at the turn of the century that, on the one hand, turned against the stylistic variety of eclecticism and, on the other hand, perceived the progressive modernization of society and its thoroughgoing mechanization of the everyday world to be a menace to cultivated ways of living and their natural and cultural context. The latter movements were less interested in aesthetic questions, for they sought to resist, above all, iron's "brutal intrusion into life and nature." Understandably, iron buildings thus also became the target of this peculiar mixture of a will to reform and loyalty to tradition.

"The magic of a solitary mountain landscape," as it is referred to in an appeal of 1904 of the Deutscher Bund Heimatschutz (German conservation federation), "is destroyed by obtrusive buildings. Iron bridges with their ugly, landscape-defacing forms span our watercourses, even where every demand of purpose would have been satisfied with simple stone and wooden bridges."[99] But even the Deutscher Bund Heimatschutz, a movement that first found a foothold in Germany and shortly afterward in Austria and Switzerland, was realistic enough not to advocate the total renunciation of the constructions of engineers. Rather, the Heimatschutz came out in favor of a "gentle" use of new technologies in architecture, that is, above all in the sense of their harmonious integration into nature. Beginning in 1913, the business manager of the Heimatschutz, Werner Lindner, made these demands known; ten years later he even wrote a book on the same theme with the title *Die Ingenieurbauten in ihrer guten Gestaltung* (Good design in constructions by engineers).[100]

"Engineering Aesthetics"

The preservation tendencies of the Heimatschutz, which in architecture occasionally found expression in a plea for a national or vernacular style, presented the later biographer of Otto Wagner, Joseph August Lux, with the overt justification for a comprehensive defense of iron architecture, which he published in his book *Ingenieur-Aesthetik* (Engineering aesthetics).[101] Just like Lucae or Hofmann before him, Lux was of the opinion that a new era had been entered, one in which developed technology was the most important cultural drive. The difference now, however, was that Lux belonged to a generation that was in a position to think this assumption through to its most radical consequences; this meant nothing less than a break with the traditional aesthetic notions and the positing of an architectural conception that was deployed as a polar opposite to the doctrines of historical architecture. In the 1870s Lucae could only vaguely sketch the broad outline of this development. He could not foresee the sharpness of the conflict.

Only one year after Filippo Tommaso Marinetti, in 1909, had described the beauty of Nike of Samothrace coming to a wretched end in a collision with a roaring automobile,[102] it was the turn of another monument of the classical cultural heritage. "We cannot help it," wrote Lux in the introduction to his book, "that nowadays, from early childhood onward, technology is more important than Plato." Lux explained why this is so a little later on with the statement that "two worlds are sharply contrasted."[103] The world of technology that Lux supported arose independently, even in opposition to any tradition, in the sense of interrupting historical lines of continuity. For half a century theorists had been grappling with the idea of making a start at the "artistic transfiguration" of technical form, in order — as one hoped — to save the inherited ideal of beauty from the avalanche of advancing technology. To Lux, this idea had become obsolete.

> Should technology, which is completely a child of our time, now wear the old fashions and tailor the old, bygone styles to its youthful, gigantic body? Should it dress like a great-grandmother? It is a healthy instinct to reject altogether this retrogression toward old motifs. *Artistic form must be discovered* anew from *the new elements.* This is the problem on which we are all working.[104]

Lux explicitly presented the demand for a "new architecture," whose seeds "resided in the technical field." Its forms would be borne by a new aesthetics; its name would be "engineering aesthetics." The paradigm to which it had to conform consisted of the products of modern machine technology: "Our culture is not reflected in the architecture but rather in the vehicles, in modern transportation technology. If we inquire about the style of our time, here we have it."[105]

In positing the imperative of the new style, Lux echoed the ideas of the Deutscher Werkbund, of which he was a member until 1908. *Sachlichkeit* and purposive perfection were thus the leading concepts or, in his words, the "supreme principles." Lux again asked whether purposiveness and beauty were mutually conditioned. He saw their harmony as the realization not only of an aesthetic principle but even more that of an "ethical principle." Truthfulness and solidity release aesthetic feelings, he thought, and the more purely and completely they were expressed, the more satisfactory would be their effect. Lies and masquerades were in this sense rejected; they were even seen as hateful. It is precisely these principles that Lux projected onto the new architecture of iron, "which impressed the modern time with its distinct stamp of style." Iron was "style-forming," whereas stone killed every new formal idea. The architecture of iron was "unconditional," and the personality that embodied the new spirit in architecture was not that of the architect but that of the engineer. Lux affirmed that "the engineer is the real architect of the modern age," for "not architects but engineers have discovered the language of iron."[106]

Even though Lux viewed *Sachlichkeit,* purposiveness, and rational construction as the essential driving forces of the radical break with the tradition of architecture, on the one hand, and of the emergence of the modern, unconditional form, on the other hand, he did not want to break definitively with the idea of the creative act or with the notion of genius that belonged to idealistic aesthetics. Moreover, he was not convinced that the conditions of beauty were set principally by technical form: "Thus we should not give credence to the false view that technical genius works without regard to aesthetics or, if you will, artistic appearance. Many structures and machines owe their form not to mere mathematical calculation but to a certain kind of experience, which offers the design of certain parts to the sense of form."[107] This sense of form admittedly has nothing at all to do with any "stylistic-historical influence" but is the result, as Lux timidly notes, of a certain "presentiment of form." In conceiving technical form, the engineer proceeds approximatively, and precisely this fact provides the channels through which a "sense for a certain exterior harmony" can prevail.

Lux admittedly did not go so far as to propose a grammar of engineered form, but he nevertheless roughly sketched the formal characteristics of iron construction. In the process, he reinterpreted in a positive way all the reproaches that had been made against the works of engineers during sixty years of debate. Thus the antimonumental effect of iron construction became one of its greatest virtues, the inclination toward "dematerialization" its supreme law, and so on. In contrast to stone architecture, the goal of which was the creation of spatial enclosures, iron construction was a matter "of mere linear contours, of the fleshless skeleton or the open frame, in short, the support that transmits the energies and represents lines of force." Like Hofmann before him, Lux emphasized the linearity of iron construction as the essential mark of its aesthetic effect. Through this arose "new spatial images." He declined to expand on this statement, but he stressed that the perception of the beauty of iron construction demanded a "new eye."

Lux also discussed ferroconcrete—warning against it rather than approving of it. He saw the dematerialization produced by iron construction as jeopardized by the fact that ferroconcrete reintroduced corporeality into architecture and thus could lead to the temptation of wanting to bring about, although with other means, "a new era of stone building."

Iron: "Style-Forming" and "Style-Inhibiting"

Lux's *Ingenieur-Aesthetik* was a polemical tract, the manifestation of a conviction rather than an argumentative effort. What Lux did not achieve, Alfred Gotthold Meyer, an art historian and professor at the Königliche Technische Hochschule (Royal technical institute) in Berlin-Charlottenburg, had sought to accomplish a few years earlier. His book *Eisenbauten* (Iron buildings),[108] which appeared three years after his death and was planned as the first part of a stylistic-historical panorama of the nineteenth century, remained incomplete (a supplementary final chapter was added by Tettau).

Meyer's enterprise was twofold. First, he attempted to explain theoretically the special features of iron construction, accompanied by reflections on its aesthetics. Second, he intended to deduce historically the beginnings of iron construction and its further development. The fulfillment of this last task, which Tettau undertook in the added "fourth section" of the volume, entitled "Kunstformen" (Art-forms), was the least successful. It is a listing of realized iron constructions in cast iron and rolled iron, without any grand theoretical claims and intentions.

Regarding the origin of modern iron construction, Meyer referred to the debate at the end of the eighteenth century concerning the nature of architecture. It had divided minds into two camps: those who held that architecture was a "*naturalisation scientifique,*" based upon mathematical rules and calculation, and those who saw it as a pure creation of artistic genius. According to Meyer, the shift occurred with Jean-Baptiste Rondelet, who "treated the structural calculation for the first time as an essential part of the discipline of construction." This opened the path to iron's use as a building material, for "the nature of iron building is based on 'rational construction,' by which it is able to resist stress and strain with far less material volume than any other material."[109] Profiting from a new mode of building based on calculation, iron construction had contributed to its further development up to the point of perfection, that is, from calculation into formal vision. "It stands . . . so to speak, just at the end of that synthetic path, in which the problems of mechanics are being transposed from the realm of arithmetic operations and algebraic formulae into the vision of graphic forms."[110]

Thus we may speak of an aesthetics of iron construction in two respects. Since the purpose of calculation is the building, an "inner vision" is required that is able to grasp the finished form visually before its erection, a "pictorial 'thinking,'" as Meyer calls it. "Such 'calculation' is then a disposing of forces, in which what is sought is *no* longer what is numerically and technically 'rational,' but rather what has the formally favorable *effect.*"[111] The "intellectual activity of constructing" and the "sensuous imagination" are joined; the intellect even surrenders to imagination.

On the other hand, iron construction as a finished, built artifact affects the observer in the immediate sense of an *aisthesis* (sensory perception). Unfortunately, Meyer very soon abandoned the epistemologically risky realm of "pictorial thinking" in order to examine the aesthetics of iron construction exclusively in the realm of perception. This posed the question of the "style-forming" or "style-inhibiting" aspects of iron architecture.

Meyer was of the opinion, as was August Schmarsow,[112] that architecture was, above all, an art of space, and that the essential task of architectural creation was the design of space. The "new spatial value" that iron construction yielded was therefore the foremost aspect of his consideration. He examined the Crystal Palace, built for the London World Exhibition of 1851, with this in mind. Meyer shared Lucae's enthusiasm and repeated the latter's characterization of the building as a "piece of sculptured atmosphere." He spoke of the "limitless space," of the "bright space" that was brought about by the shadowless, glazed architecture flooded with light, which was still connected — through its structural linearity and Owen Jones's color treatment — to the "realm of measures and space" (perhaps in the sense of spatial limitations). These last elements — line and color — were, so to speak, the stylistic support for the building. Although Meyer recognized their artistic value, he maintained that they were insufficient for "space formation." The Crystal Palace's latticework was capable only of mere "spatial enclosure" and "spatial cover."

According to Meyer, the "new width" was the second aspect of the aesthetics of iron construction, which he described by means of the Galerie des Machines at the Paris World Exhibition of 1889. Here classical tectonics, which had always presented the problem of the relation between "support and load," had been overcome. The hall of pillars and columns, in which the design of architectural space is based on spatial division, had now become obsolete as the main form of spatial design in architecture. The hierarchy of the axes of expansion of the classical spatial system had been stood on its head. Height was no longer the dominant axis, followed by depth, the direction of free movement. Breadth now played the critical role in the characterization of space. This unified space — *Einraum,* as Meyer called it — had been brought about by the self-supporting roof, which became possible only by means of iron construction.

The "decisive power of the line" became Meyer's third aesthetic characteristic of iron construction, the "new height" that it produced. Meyer investigated this aspect in the Eiffel Tower, which he characterized as "a landmark in human history." He again took up Semper's argument about the "invisible architecture" of iron and objected that the three-hundred-meter-high Eiffel Tower was the *most visible* structure in Paris. "Invisibility" was not its characteristic, but rather transparency, and he maintained that this arose from the reduction of mass to surface and of surface to line. In judging it, the historical concept of style, which had developed from "other materials and other constructional achievements," likewise failed. Although Meyer stressed the Eiffel Tower's "charm of the stupendous" and even its mysterious effect, he noted that these judgments had "nothing to do . . . with aesthetic pleasure." He even criticized the linear labyrinth in some parts of the structure as style-inhibiting. What Meyer experienced as nonaesthetic, he actually described in terms of an aesthetics of the sublime. His own aesthetic view was, however, so bound by the idea of beauty that he could not exploit the theoretical opportunity that he himself had opened up.

The fourth and last part of the aesthetic tetraptychon on iron construction carried the title "Neue Linien" (New lines). Here the emphasis was on bridge construction and the question of whether it should be classed as architecture. Meyer thought that bridges did not shape space but that they were "without doubt an architectural task." The "power of the line" expressed itself here in the sense of the contour of the structure, and it was at the same time its most important aesthetic component.

Meyer's concept of architecture started with an aesthetics of space in the sense of creating spatial divisions and enclosures. But he had to relinquish it piece by piece in the course of his presentation of iron construction. The expectation that this process could generate a new conception of space remained unfulfilled. The question of balance between "style-forming" and "style-inhibiting" elements of the (no longer) new construction was not resolved. He was thoroughly convinced that with the help of iron the birth of a new architecture was taking place, an architecture that could find acceptance even if through a process of "habituation to form" (*Formgewöhnung*). But he also maintained that "the 'building-art' arising in this way would never be able to drive out and replace monumental style-architecture, as some all-too-ardent prophets of iron construction were claiming."[113]

An Almost Aesthetic Pleasure

The idea of "pictorial thinking" (Meyer) or "sense of beauty" (Lang) — later described by Muthesius as "aesthetic subconsciousness," a process that was active at the time technical form was being invented — was further developed by the art historian and cultural critic Karl Scheffler. His book *Moderne Baukunst* (Modern architecture) dealt with contemporary architectural questions and appeared in the same year as Meyer's *Eisenbauten.*[114] Scheffler did not dispute that moments of imagination contributed to the phase of inventing technical form. He even went so far as to claim that precisely for this reason technical invention went beyond profane considerations of usefulness and as such belonged "to the domain of that higher purposiveness based on knowledge," that is, to the domain in which beautiful stylistic forms reside. It is quite a different case with construction, which was directed toward "profane material necessity," toward the functioning of building members. Construction was admittedly the point of departure for form; it created the basis for it, even though it was no "free creation of form." And thus Scheffler could aphoristically maintain that "construction is not art,"[115] for art was independent from constraints of matter.

"The artist dematerializes matter,"[116] Scheffler characteristically noted. Yet he thought this "dematerialization" was only metaphorical, that the artist "substitutes the symbol of necessity for necessity itself." Otherwise, the act of liberating matter would presuppose the existence of matter, that is, in the sense of the presence of mass from which the artist could shape the art-form. The conclusion was thus evident and history had, according to Scheffler, repeatedly shown "that there has never been a monumental architecture that has not made use of stone."[117]

This view at the same time determined the author's position on iron. This material was, in a much truer sense than any other that preceded it, a constructional material, "and thus it excludes the free art-form even more."[118] In fact, every attempt to win an art-form for it had foundered. How could it be otherwise for a building material that was "completely lacking in plasticity?" As Scheffler noted, "the line means nothing in architecture, mass everything."[119] Nevertheless, he was prepared to grant a certain aesthetic value to engineering constructions. This was expressed in the features of a "heroic monumentality," in a "primitively refined magnificence." Scheffler characterized the respective sensations as "an almost aesthetic pleasure" and thus unconsciously found himself, just like Meyer, close to the aesthetics of the sublime. Far from disputing iron construction's right to exist, Scheffler even spoke of an iron style: "The iron structure has a type of beauty of the second degree, a bound constructional beauty and a particular idea of style. The result: the iron style."[120] He nonetheless rigorously opposed the attempts at a subsequent "aestheticization" of iron: "It is madness to want to salvage higher aesthetics by associating historical art-forms with iron constructions."[121]

Scheffler's standpoint implied a stylistic duality that depended both on the use of stone (fit for art) and on the pure constructional material of iron. These two styles — or "form worlds" — were irreconcilable, and any attempt to combine them was doomed to failure. In a certain respect, this view of things implied a radicalism of appearance that bound Scheffler not with the defenders of stone architecture but rather with Lux and the later avant-garde: stone, which during the entire debate represented the normative concept of architecture, and iron, to which the role of the embodiment of the modern zeitgeist was attributed, were — according to Scheffler — not complementary but incompatible. However, the author did not use this incompatibility to establish a historical break, but rather, as previously mentioned, a duality of styles. Scheffler did not divulge what this dualism might mean for architecture's evolutionary perspective, and he concluded his pertinent observations with the following enigmatic lines:

> If one day we were to have our own grand architecture, the characteristics [of iron structures] would certainly recall in more ways than one the severe, almost Gothic seriousness of engineers. A long excursion is still needed before lasting results can be achieved; but iron will better serve the architecture of the future the less it claims to be an artistic material.[122]

A Powerful Force in Simple Form

On 15 January 1908 the Königliche Akademie des Bauwesens (Royal academy of building) in Berlin arranged a competition open to "subjects of the German Reich" for the best "treatise on the artistic design of iron construction in the field of architecture and engineering." The submissions were to include verbal and pictorial presentations of previous achievements in this field, a critical evaluation of them, and the prospect of the aesthetic possibilities of iron construction. The results of the competition were disappointing: in all, only five contributions were received by the jury, which consisted of the president and six members of the academy (three architects and three engineers). The committee found none of the submissions worthy of the first prize and therefore refrained from awarding one.

The reason for this fiasco is that by the end of the first decade of the twentieth century this subject had largely been exhausted. During more than fifty years of debate on the aesthetics of iron construction, all the essential arguments had already been made and all the main positions articulated. There was hardly anything new to add. This was confirmed, in fact, by two revised essays, both of which won second prize and were published by the academy.

A passage in the first essay, written by the royal building commissioner in Strasbourg, the engineer Dr. Hermann Jordan, merits special attention.[123] It is in the section with the subtitle "Allgemeine Betrachtungen über die ästhetische Wirkung von Eisenkonstruktionen" (General observations on the aesthetic effect of iron constructions).[124] According to Jordan, this subject extended beyond the narrow boundaries of this branch of knowledge and could only be discussed within the framework of a discipline that possessed an appropriate theoretical instrument for such a question: philosophical aesthetics. The author criticized the notion that the aesthetic pleasure of viewing a building depended on the degree of its purposiveness and on understanding the function of its parts. Muthesius provided him with the argument: functional purpose was not the only critical aspect of the modern designer's work; the unconscious drive toward pleasing forms should also be at work.[125] Jordan expanded this position by maintaining that between the basic structural idea and the design there is a phase of "continuous selection between different possibilities." To this he added, "Thus scientific considerations and calculations are often of little help"[126] — implying that the constructional form cannot be accounted for simply by arguments on the level of the rationality of purpose. But he relativized even the thesis of "pleasing forms," in that he asserted that the aesthetic effects of iron construction did not necessarily mean that it was also beautiful. "Aesthetically effective" is a much more general concept than "beautiful."[127]

Jordan seems to have had his doubts even about the harmonious effect (as a condition of beauty) of iron construction. Harmony was "unity in multiplicity," he noted, and from this position he compared the machine form with the iron structure. One could well claim that a machine or a vehicle was harmonious because its design depended on the harmony of its various functional parts. The "unity in multiplicity" would be realized here, just as it was in nature, even if on a lower plane. Hence there was the aesthetic effect of locomotives, vehicles, bicycles, and ocean liners. It was completely different with iron construction. For there one had no multitude of functional parts but rather very simple, clearly visible functions that could be met by a small number of structural elements. Nevertheless, one could also speak of an "aesthetic effect" of iron construction that was in no way inferior to that of machines. Jordan solved the apparent puzzle of this position with the help of Karl Groos's *Aesthetik*.

> In order better to explain this effect, I must again return to the book by Groos that has already been mentioned several times. Among the various aesthetic modifications, Groos also explains the concept of the sublime. His investigations led him to the conclusion that the sublime required, on the one hand, powerful dimensions exceeding the average, and, on the other hand, simple forms. "*The sublime is a powerful force in simple form.*"[128]

Uncertain of his own discovery, Jordan did not dare to attribute fully the characteristic of the sublime to iron construction. Thus he spoke of the Eiffel Tower, the "American suspension bridges," and the large railroad terminals as works that were "more or less 'a powerful force in simple form.'" Nevertheless, with respect to large iron construction Jordan was the first to abandon the aesthetics of the beautiful. He opened the perspective of the aesthetics of the sublime and named it as such. This occurred inconspicuously, one might say incidentally, in an otherwise very mediocre text.

Ferroconcrete

In 1911 E. von Mecenseffy's *Die künstlerische Gestaltung der Eisenbetonbauten* (The artistic design of ferroconcrete buildings) was published as a supplementary volume to Fritz von Emperger's multivolume *Handbuch für Eisenbetonbau* (Handbook of ferroconcrete building).[129] Mecenseffy approached his subject with extreme care, or rather, in an exploratory manner, and the significance of his book lies less in the information conveyed than in the fact that it was one of the few works that attempted to take a position on this issue. It was, at least, the first thorough investigation of ferroconcrete to be written systematically.

From the end of the 1880s onward, and above all in the first and second decades of the twentieth century, there was considerable discussion of this topic in architectural journals. Yet in a surprising way, this very extensive architectural literature on concrete and ferroconcrete was limited to the technical aspects of application and hardly touched upon aesthetic problems. The reason for this might be found in the conditions surrounding the introduction of concrete as sketched by Mecenseffy. He thought that the advent of this material in architecture had encountered far less resistance than iron and gave two reasons for this. The first was the general condition of architecture at the time, which was characterized by "a certain feeling of unease [*Katzenjammerstimmung*] toward the inherited world of forms," and the second was to be found in the properties of the new material itself, namely, concrete's ability to adapt "to any possible formal language." As he noted, "this coincidence appears to have been the reason that ferroconcrete was approached in a more unbiased way than, for example, its immediate predecessor among the new building materials — iron — the appropriate form for which one still struggles after heatedly today."[130] Mecenseffy's personal view was that the promise of a new architectural style linked to the use of iron had not been redeemed: "Today we still experience such forms as ugly and strange."[131] He noted that as far as the aesthetics of ferroconcrete itself were concerned, the process of producing concrete buildings (wherein a formless material is stamped into forms) "imposed on the architect an extensive self-denial" because the forms permitted only "fairly meager geometrical surfaces and bodies."

The same view was represented a few years earlier by the Austrian architect of the courts of law in Salzburg and Brno, Alexander von Wielemans, at the seventh International Congress of Architects, 1906.[132] It actually seems that Mecenseffy borrowed his ideas about the aesthetic design of ferroconcrete largely from Wielemans's publications. In his excellent study *Concrete,* 1959, Peter Collins rightly emphasized Wielemans's closing remarks on the question of style. Wielemans noted that he was not preoccupied first and foremost with the search for a new style, for a style arose in a fairly natural way out of the methods of production that the material itself dictated. Nevertheless, in his own buildings Wielemans could not completely do without decorative elements applied to the concrete shell of the house. This seems, in turn, to confirm Collins's thesis: "Throughout the whole period, the originality being sought in concrete forms tended to be the originality of artistic inspiration, rather than the originality conferred by new structural methods."[133]

Wielemans delivered the same lecture at the next international congress, which took place in Vienna in 1908. Two years later, the annual meeting of the Deutscher Werkbund convened in Berlin to address, among other things, the topic "material and style." In his relevant remarks, the first president of the Werkbund, Theodor Fischer, rejected the view that the use of the new material alone might lead to new forms, and he drew attention to the related design task facing the architect: only his achievement could transform concrete into an art-form. Karl Ernst Osthaus expressed himself in a similar fashion, seeing the possibility of a new style emerging from the interplay of the new technical possibilities and the architect's imagination.[134] In addition to their influence on the internal debate of the Deutscher Werkbund, these observations possess a particular significance, for they lead smoothly to the intellectual edifice of that architect whose work became the focus of Giedion's remarks on ferroconcrete in *Bauen in Frankreich* — Le Corbusier. As a member of the audience at the Berlin conference, Le Corbusier not only carefully recorded the local course of events but also integrated the ideas articulated there into his theory, as can be seen — at the latest — in his book of 1923, *Vers une architecture.*[135]

"Architecture?"

Shortly before the outbreak of World War I, all the essential arguments for and against the use of iron and ferroconcrete in architecture were thus on the table. When Giedion in 1928 (in the context of "new architecture") addressed the question of the relation of architecture to technology and sought to answer the question historically in his book *Bauen in Frankreich,* the themes upon which his argument developed had almost been forgotten. As Peter Meyer noted, the subject, "despite its topicality, had almost disappeared."[136]

Forgotten, above all, was the great debate over "dematerialization," which the shock of iron construction had provoked. During the year before *Bauen in Frankreich* was published, one of the staunchest advocates of the new architecture, Walter Curt Behrendt, described the "buildings in the new style," above all in terms of their corporeal qualities. They revealed their "common characteristics" as "works of simple, austere form and clear construction; with flat, smooth walls, completely flat roofs, and rectilinear contours. The articulation of the building bodies, as a rule, was produced only by a more or less lively stepping of the building masses and by the distribution of windows and openings on the wall surface."[137] In this connection Behrendt thought that the new technical facts, derived from the use of "new building materials such as iron, concrete, and glass," comprised one of the essential factors of the "new style," which he logically referred to as the "material style" (*Materialstil*).[138]

But corporeality was not to Giedion's liking. He therefore gave new life to the large iron constructions of the nineteenth century, which were used for the completely new building tasks of the time: exhibition halls, railroad stations, and large department stores. He made them the elements of a second, parallel, and alternative history of architecture — in a specific overturning of the discipline's traditions that pointed directly to the Modernism of the twentieth century. He selected France as the field of his investigations even though his theory had profited mainly from the debate in Germany, which had been going on among architects for eight decades. The concepts he used in describing and interpreting iron construction, his thoughts regarding the role of the architect and the engineer, his ideas on the interrelationships between a visionary conception and a rational calculation of structures — all provide unmistakable and undeniable evidence of his debt to that debate. Even the imbalance between the first part of the book (in which he conceptually grasped the aesthetics of the new architecture by reflecting on iron construction) and the second part (in which he applied this aesthetics to buildings in ferroconcrete) was a clear reflection of the German debate.

That kind of connection between building in iron (the creations of the engineers of the nineteenth century) and building in ferroconcrete (the works of architects from Auguste Perret via Tony Garnier to Le Corbusier and "the next generation") is the new element in *Bauen in Frankreich*. This fact becomes especially clear if one compares it to another book, which appeared at almost the same time as Giedion's and which also had as its object the architectural possibilities of ferroconcrete: Julius Vischer and Ludwig Hilberseimer's *Beton als Gestalter* (Concrete as shaper of form). The introductory text by Hilberseimer admittedly established a connection between the iron skeleton and the ferroconcrete skeleton, as when he referred to them as two methods of scaffolding construction. Yet, when it comes to the architectural design of ferroconcrete, this connection appears largely without background, as a type of parthenogenesis.[139]

Giedion's historical operation is not without problems. Aesthetically, he situates works that were profoundly different in their corporeal presence in a single line of tradition. These include the skeletal engineering structures and the cubic, corporeal productions of ferroconcrete, above all those of Le Corbusier. The basic incompatibility and aesthetic incommensurability can be neutralized only when one assumes the perspective of the observer as a point of departure for the events. And this is precisely what Giedion does.

Equipped with the visual apparatus of modern man — formed and educated, above all, through the study of Cubism and Neoplasticism — he wanders through his objects in cinematographic fashion.[140] Thus it turns out that for purposes of their description and interpretation he can uniformly apply the conceptual and theoretical instruments that he had developed by recording nineteenth-century iron construction — as a comparison of two of the figure captions from *Bauen in Frankreich* will reveal:

Arnodin. Pont transbordeur, Marseilles. 1905
[This structure] cannot be excluded from the urban image, whose fantastic crown it denotes. But its interplay with the city is neither "spatial" nor "plastic." It engenders floating relations and interpenetrations. The boundaries of architecture are blurred.[141]

Le Corbusier. Pessac housing settlement. 1925
Corbusier's houses are neither spatial nor plastic: air flows through them! Air becomes a constituent factor! Neither space nor plastic form counts, only RELATION and INTERPENETRATION! There is only a single, indivisible space. The shells fall away between interior and exterior.[142]

Giedion's procedure resembled nothing less than a theoretical somersault. It was only justified if one completely overlooked and ignored the corporeal attributes of architecture. That such a method had to lead to a new concept of architecture was a component of the strategy Giedion pursued in *Bauen in Frankreich*. He was certainly not the first to demand that the discipline withdraw from its normative boundaries. The discussion about the "new way of building," or the "new style," had — since Bötticher — touched the core of this problem. It spread itself like a film over all efforts, not only on a theoretical but also on a practical level, and finally led to modern architecture. Hence, Behrendt did not stand alone when he wrote in his aforementioned book, "Therefore no alternative remains but *to begin anew,* to give up the old concepts of form that have become invalid and proceed in our own independent way by *constructing, forming, designing.*"[143]

Giedion's way of seeing led him one important step further. He not only challenged the appropriateness of the inherited concepts of form, he also questioned the legitimacy of the architectural discipline itself. This element constitutes, so to speak, the relevance of his book to the time when it was published. It is best seen in a single phrase in the book, which, under the chapter heading "Architecture?" is presented almost as a conclusion to the history of the book: "It seems doubtful whether the limited concept of 'architecture' will indeed endure."[144] This expresses not only the "radicality of knowledge" that was transmitted by *Bauen in Frankreich* (as Walter Benjamin thought when he read it) but much more a radicality of attitude, that is, a "radicality of conviction."

II.
History of the Book

In a letter of 6 October 1926 Giedion offered Professor Georg Biermann, editor of the journal *Cicerone,* a series of essays on the new architecture:[145] "You shall not be lacking architectural essays from me," he wrote.[146] A proposed text on Le Corbusier would thus be expanded in order to include the whole of the "situation of French architecture." At the same time Giedion informed him that he was at work on a book. "I am working on my book on the new architecture, which will perhaps be a rather fundamental text, if my intensity percolates evenly through all parts of it."[147] In fact, a few months later three articles in a series, "Zur Situation der französischen Architektur" (On the situation of French architecture), appeared in *Cicerone:* the first in January 1927, the last in May 1927.[148] With the first issue Giedion became the "independent supervisor" of the part of *Cicerone* dedicated to modern architecture.

Sometime between October 1926 and March 1927 Giedion must have proposed to the editor of the journal that his material on French architecture be published as a book. At least Georg Biermann, in a letter of 16 March 1927, returned to this proposal after praising Giedion's contributions to the journal and speaking of their "eminent significance": "I will return to the idea of publishing a book after Taut's book has appeared.[149] . . . If it is a success, as I certainly hope, I would gladly take up your publishing idea and make a similar second volume from your writings on the new French architecture."[150] Although he received no definite commitment from Biermann, Giedion tackled his publication proposal head on.

Thus, on 24 April 1927 Giedion contacted Hans Jenny-Dürst, professor of structural engineering at the ETH in Zurich, to request that he correct any technical inadequacies in his text. This must refer to an early version of the manuscript. One year later, on 27 April 1928, Giedion reported in a letter to Biermann a discussion "with the structural engineer at the Technische

Hochschule" that had taken place that day. "Since he is a European authority, it will be advantageous, especially when we are confronted with the reaction, if we could, even belatedly mention his contribution."[151] As the reader of *Bauen in Frankreich* learns from a comment in the Preliminary Remark, the resulting corrections could not always be fully taken into account, "due to the advanced stage of production" of the book.

In June 1927 a provisional title for the book, *Neues Bauen in Frankreich* (New architecture in France), was chosen (as a letter from Biermann dated 23 June shows). Biermann, who in the meantime had apparently decided to publish the book, also held out the prospect of a French edition, published jointly with Editions Crès in Paris. Yet it appears from a letter of Giedion's to Biermann of 15 August 1927 that there was still no contract for the German edition. Giedion asked for — and this was later accepted by the publisher — 12 percent royalties, with the amount for the first one thousand copies to be paid upon publication. He left the date of publication open, for "a skillfully done work of propaganda [is] more important than a precipitate publication."[152] But he remarked that the book should be on the market no later than the beginning of 1928.

Giedion spent some weeks in Paris in late October and early November 1927 collecting material for the book. At the end of November, he was still sending letters from Zurich to various addresses in France (Diréction des Chemins de Fer du Nord, Printemps Department Store, Bibliothèque des Arts et des Métiers), above all inquiring about photographic materials that he needed for the book. On 19 November 1927 Biermann returned to the question of the title, a question that was to haunt the book for some time even after its publication. According to Biermann, "the new architecture in France should, in any case, not be missing from the title, however much your real emphasis is on ferroconcrete."[153] On the matter of the typographical design of the book, Biermann wrote:

> As regards the artistic production, we should try to make do without an artist this time, for the public has had more than enough of that in the last two Taut books. Simplicity and clarity seem to me to be the style that the time needs, even in book architecture. I believe that as far as the technical quality of the book is concerned, we can in this instance guarantee the indicated direction without the help of an artist, which under the circumstances could be very dangerous.[154]

VORTRÄGE DER STAATLICHEN KUNSTBIBLIOTHEK
IM HÖRSAAL / PRINZ-ALBRECHT-STRASSE 7A / HOF

NEUES BAUEN

7 VORTRÄGE / MONTAG, ABENDS 8 UHR / ANFANG: 30. JAN. 1928

1.
MONTAG, DEN 30. JANUAR
ARCH. C. VAN EESTEREN, HAAG
STÄDTEBAU

2.
MONTAG, DEN 6. FEBRUAR
Dr. S. GIEDION, ZÜRICH
EISEN, EISENBETON, BAUEN IN FRANKREICH

3.
MONTAG, DEN 13. FEBRUAR
PROF. HENRY VAN DE VELDE, BRÜSSEL
WARUM IMMER NEUES?

4.
MONTAG, DEN 20. FEBRUAR
ARCH. ERICH MENDELSOHN, BERLIN
RUSSLAND — AMERIKA,
EIN ARCHITEKTONISCHER QUERSCHNITT

5.
MONTAG, DEN 27. FEBRUAR
ARCH. MIES VAN DER ROHE, BERLIN
DIE VORAUSSETZUNGEN
BAUKÜNSTLERISCHEN SCHAFFENS

6.
MONTAG, DEN 5. MÄRZ
ARCH. MARTIN MÄCHLER, BERLIN
DAS CITY-PROBLEM

7.
MONTAG, DEN 12. MÄRZ
ARCH. I. I. P. OUD, ROTTERDAM
VON MODERNER MALEREI UND HEUTIGER
TECHNIK ZUR NEUEN ARCHITEKTUR

DIE VORTRÄGE WERDEN DURCH LICHTBILDER ERLÄUTERT

EINTRITTSKARTEN FÜR DIE GANZE REIHE 10 MARK / FÜR EINZELVORTRÄGE 2 MARK / IM LESESAAL DER STAATLICHEN KUNSTBIBLIOTHEK WOCHENTÄGL. VON 10—10 UHR / SOWIE AN DER ABENDKASSE

FÜR MITGLIEDER DES FREUNDESKREISES HALBE PREISE.

„Bukwa", Berlin C 2. Breite Straße 8—9

Giedion began the "propaganda" for his book at the beginning of 1928. On 6 February he was in Berlin speaking on "Eisen, Eisenbeton, Bauen in Frankreich" (Iron, ferroconcrete, building in France), in a series of lectures at the Staatliche Kunstbibliothek in which other prominent representatives of the new architecture also took part (fig. II). The new architectural tasks of the nineteenth century — railroad stations, department stores, exhibition halls — were central to his remarks, and they were discussed from the point of view of the aesthetic possibilities of using the new materials and structural methods. The contents of his lecture were thus within the framework of his forthcoming book. As Alfred Gellhorn commented, "The lecturer was, on the whole, the kind of speaker one may meet only once in a lifetime. His highly succinct manner of speaking, presenting totalities and penetrating the essence with only three words, signified something extraordinary."[155]

In view of his subsequent publication, the title Giedion chose for his Berlin lecture is revealing. Building with the new materials iron and ferroconcrete appeared as the first part of the title, to which was added the reference to the country that served as the field for his investigation, France. Furthermore, Giedion used the word *Bauen* (building), not the notion *neue Architektur* (new architecture), which Biermann had proposed as part of a title for the book. That this corresponded to Giedion's idea of a book title is borne out by an undated typescript whose accompanying instructions indicate that it was intended for the typesetter. It repeats the title of the Berlin lecture, but gives as an additional component of the main title "CONSTRUCTION BECOMES DESIGN."[156]

A few days after his Berlin engagement, Giedion was in Hannover. In a lecture to Alexander Dorner's Kestner Society (fig. III), Giedion drew an arc from the first iron constructions of the nineteenth century to the work of Le Corbusier and established himself as "an enthusiastic apologist for the most recent architectural intentions." According to one attendee, the lecture was accompanied by "a wealth of largely unknown slides."[157] On 1 June Giedion spoke on the same topic at the University of Zurich. The publication of his book was announced on this occasion.[158]

Giedion wrote to Robert Mallet-Stevens on 8 March 1928 that the book was at the typesetter's. In April Giedion was in close contact with the publisher concerning the correcting of proofs, and this process lasted until the end of May. He placed great importance on this matter, for "the material of the book is quite delicate, and I must be painstakingly conscientious about how I express myself on such a controversial subject."[159] Neverthe-

Fig. II. Announcement of a lecture series at the Staatliche Kunstbibliothek in Berlin. This event marked Giedion's first lecture on the subject of his book *Building in France*. Courtesy Archiv S. Giedion, Institut für Geschichte und Theorie der Architektur, ETH-Hönggerberg, Zurich.

IM HAUSE DER

KESTNER-GESELLSCHAFT E.V.

KÖNIGSTR. 8 SPRICHT

DONNERSTAG

DEN 9. FEBRUAR • ABENDS 8 UHR

DR. S. GIEDION-ZÜRICH

BAUEN

IN FRANKREICH

IM 19. UND 20. JAHRH.

ZUR TRADITION DES NEUEN BAUENS • MIT

140

BISHER NICHT VERÖFFENTLICHTEN LICHTBILDERN FRANZÖSISCHER BAUTEN

EINTRITT 2 MK. • MITGLIEDER 1.50 • SCHULER 1 MK.

less, there were continual disagreements, for instance, when Giedion noticed in his first galleys that the question mark in the chapter title "Architecture?" was missing. "This gives the whole chapter a different meaning," he wrote to the publisher on 25 May. "I must ask you not to undertake any changes of meaning without asking me, or else unwelcome things will happen for which I cannot be responsible."[160] The error nevertheless remained uncorrected, and the question mark only appears in the book's table of contents.[161]

Giedion was also very concerned about effective advertising. He planned preprints for large newspapers and put together lists of potential reviewers. The publisher's advertising campaign actually began in early June. A four-page prospectus was published (fig. IV) from which we learn the definitive title of the book, about which there was a fight until the very last moment.[162] Sometime between the end of January and the end of May (the exact date cannot be determined), Giedion decided on the title *Bauen in Frankreich, Eisen, Eisenbeton* (Building in France, Iron, Ferroconcrete). As an alternative, he considered reversing the order, that is, *Eisen, Eisenbeton, Bauen in Frankreich*. But he categorically rejected the rather nebulous proposals of the publisher. These wavered between the title *Neue Baukunst in Frankreich* (New architecture in France) and a formulation beginning with the word *Eisenbeton* (Ferroconcrete).[163] The final title did not actually exist until 1 June 1928, as is clear from a letter of this date written by Giedion to the publisher:

> Confirming our two telephone conversations. Following our discussion today, the cover and jacket will read:
> BUILDING IN FRANCE
> BUILDING IN IRON
> BUILDING IN FERROCONCRETE
> The inner book title [will read]:
> BUILDING IN FRANCE
> IRON
> FERROCONCRETE
> . . . The key word in the literature must be BUILDING IN FRANCE.[164]

The prospectus also gives the name of the book's designer: "Prof. Moholy-Nagy, Dessau." Exactly when Moholy-Nagy was taken on (against the publisher's original intention to avoid hiring an artist) cannot be deter-

Fig. III. Invitation to a lecture by Sigfried Giedion on the subject "Building in France in the Nineteenth and Twentieth Centuries" at the Kestner Society in Hannover on 9 February 1928. Courtesy Archiv S. Giedion, Institut für Geschichte und Theorie der Architektur, ETH-Hönggerberg,

SIGFRIED GIEDION

BAUEN IN FRANKREICH EISEN EISENBETON

130 Seiten und 139 Abbildungen
Buchausstattung Prof. Moholy-Nagy-Dessau
Preis:
Kartoniert M. 8.50
Ganzleinenband M. 10.50

KLINKHARDT & BIERMANN
VERLAG / LEIPZIG

Als Dr. Giedion vor einem Vierteljahr im Rahmen der Vorträge der staatlichen Kunstbibliothek Berlin die Grundzüge seines Buches zum erstenmal öffentlich entwickelte, hatte er einen geradezu sensationellen Erfolg, der wohl als Prognostikum für das Buch angesprochen werden dürfte. Wie keiner war er berufen, gerade dieses Buch zu schreiben, denn er ver-

mined. However, it is certain that he followed Giedion's very precise instructions about the typographic design. Giedion's contribution took the form of layout sketches from his own hand (see Appendix). The book was finally printed during the last days of May, and an edition of three thousand copies was published at the beginning of June.

Bauen in Frankreich, Bauen in Eisen, Bauen in Eisenbeton met with a mainly positive response. People were especially impressed by its catchy design, its wealth of illustrations, and its striking textual commentary. A few aspects appear to have played a special role in its reception. Thus several reviewers saw the book as a rehabilitation of the nineteenth century. "When someone like Giedion places the accents correctly," wrote C. O. Jatho, "one arrives at a fuller appreciation of the much-maligned nineteenth century."[165] Reading the book in a similar way, Karl With maintained that the new architecture with respect to "the construction problem already is a 'legitimate part' of the development of the nineteenth century."[166] The anonymous commentator from the *Dresdner Zeitung* spoke of a "new discovery of the nineteenth century in a field . . . that is generally despised: that of architecture."[167] And Alfred Gellhorn even characterized the book as a "clearance of the nineteenth century's name."[168]

A series of reviewers stressed the relevance of the topic, above all, with respect to the lack of assent still plaguing the new architecture. Gustav Stotz thus noted:

> Giedion uncovered . . . the basis of a tradition, which may be described as the only true one for our contemporary architecture, whose decisive forms are being created in iron and concrete. The book gains enormous relevance and liveliness through the way it reveals the connections and relations between the constructive efforts of the previous century and the modern efforts in the field in question.[169]

Curt Glaser wrote that Giedion's book shows that the new architecture "did not spring arbitrarily from the brain of a few artistic revolutionaries, but rather had been prepared for a long time."[170] "After reading this book," declared Peter Meyer, "no one will again be able to dismiss modern architecture as traditionless and antitraditional Bolshevism nor maintain that national nuances become smothered by constructional buildings."[171]

Yet the national aspects of the book clearly irritated many, mainly German critics: "One would have wanted to have the only lightly sketched connections to the great constructors of England and Germany, and even

Fig. IV. Page one of a four-page advertisement for *Building in France* with handwritten annotations by Sigfried Giedion. Courtesy Archiv S. Giedion, Institut für Geschichte und Theorie der Architektur, ETH-Hönggerberg, Zurich.

America, further elaborated," wrote Gellhorn in the previously mentioned review.[172] Paul Joseph Cremers did not share Giedion's optimism regarding the further development of "industrialized architecture" in France and added confidently, "Many signs indicate that Germany will take the lead."[173] On the other hand, there were also instances where *Bauen in Frankreich* became the occasion to denounce the backwardness of the German scene. This was the case with Paul Klopfer, whose main assault was aimed at the advocates of the Heimatschutz.[174]

Some critics stressed the imbalance between the two parts of the book. The chapter on iron generally came off better than that on ferroconcrete. According to Gellhorn, for instance, "the first part of the book, which is dedicated to building in iron, is the most enlightening and stimulating."[175] The reviewer for *Hoch + Tiefbau* found that the development of ferroconcrete was "only lightly touched upon."[176]

Understandably, there were a few who disapproved of *Bauen in Frankreich* as a whole. The viewpoints varied from critic to critic. Leo Adler, for instance, reproached Giedion for having fallen into the very difficulty he professed to criticize: aestheticism. Because he polemicized against stylistic romanticism, he closed his eyes to the romanticism of twentieth-century engineers, as realized, for instance, by Gropius in the Bauhaus in Dessau. Adler characterized the book's passages on Le Corbusier as "pure propaganda."[177] Herman Sörgel expressed this differently; he felt that the aesthetics that Giedion derived from the materials of iron and, in particular, ferroconcrete were unsuited for precisely the building task where they lay claim to validity in the twentieth century — residential construction. "When Giedion speaks of the demand for flushing the new house with air, inside and out, from below (!) and above, then one can just as well say that this is precisely what the house should protect us from, for otherwise we would hardly need a roof over our heads."[178]

But the general feeling remained positive. Prominent critics expressed their enthusiasm for the book. Its modernity appeared to be the main reason for its positive reception. Max Osborn thought that "this development [of modern architecture] has never been shown more concisely, more plausibly, or more obviously (particularly the captioned illustrations)."[179] Justus Bier joined him, speaking of "Giedion's bold and captivating book."[180] And, finally, Adolf Behne noted, "The book, filled with important new material, with Moholy's splendid attention to typography, belongs with those that are indispensable."[181]

Among the judgments pronounced on *Bauen in Frankreich* are those contained in personal letters to Giedion. Paul Klopfer was among the first to contact him: "It was particularly satisfying for my own deductions," he wrote on 25 June 1928, "to get assurance from you that there was a question mark behind architecture in the usual sense. I will also order the book for my school, for it holds great pedagogical value. My somewhat old-fashioned faculty members may also benefit, and I, too, for they do not always believe me."[182]

Walter Benjamin, who at Giedion's urging received a complimentary copy from the publisher, accepted it with the following enthusiastic words:

> Dear Mr. Gidion [*sic*]:
>
> When I received your book, the few passages that I read electrified me in such a way that I decided not to continue with the reading until I could get more in touch with my own related investigations than I had been, due to external circumstances, when the book arrived. In the last few days things have started to move again, and I spend hours with your book, in admiration. So far, I only know its last part. I deliberately write while I can control the excitement it has caused me. Your book presents one of those rare instances familiar to everybody: to know, before even touching something (or someone, a book, a house, a person, etc.), that this touch will turn out to be most significant. This premonition does not deceive.
>
> I am studying in your book (among so many other things that most directly concern me) the difference between radical conviction and radical knowledge that refreshes the heart. You possess the latter, and therefore you are able to illuminate, or rather to uncover, the tradition by observing the present. Hence the nobility of your work, which I admire most, next to its radicalism. I would be very glad to be able to talk with you about this. I would be grateful if you were to remember me during some stay in Berlin. If I am in Paris, in the spring, I will take the liberty of letting you know.
>
> In closing, please take with my last words my tardy thanks for *sending* me your book.
>
> With sincere compliments
> 15 February 1929
> Berlin-Grunewald
> Delbrückstraße 23
>
> Yours very sincerely,
> Walter Benjamin[183]

22475

Bilder bis an den Rand!

Eiffel: Glaswand der Pariser Ausstellung von 1878

IST DAS

NEUE

BAUEN

EINE

MODE?

LEBEN WIR

TRADITIONSLOS?

1926

AUFKLÄRUNG

ÜBER DIE GANZE

ENTSTEHUNG

DES NEUEN

BAUENS BRINGT:

gropius, bauhaus dessau

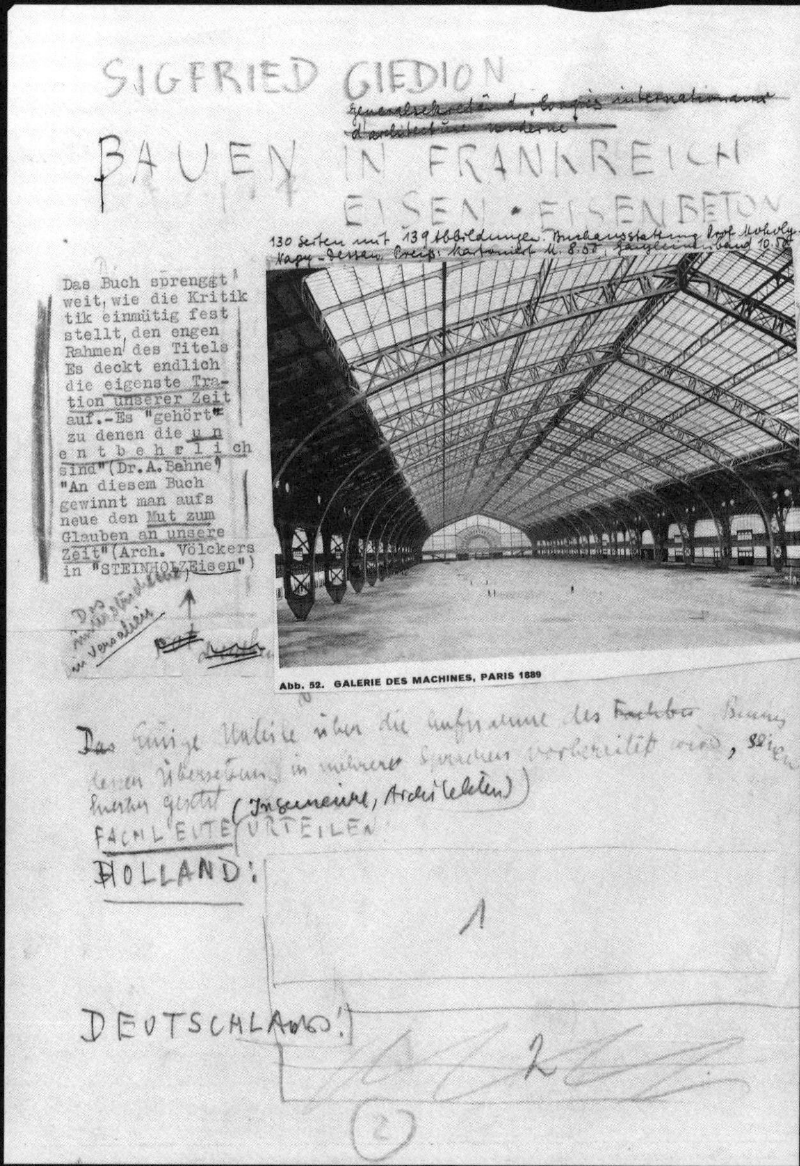

Figs. v, vi. Sigfried Giedion's design for a flyer prepared some time after the publication of *Building in France*. Giedion wanted to incorporate the critical comments on his book in this new flyer. Courtesy Archiv S. Giedion, Institut für Geschichte und Theorie der Architektur, ETH-Hönggerberg, Zurich.

Giedion also sent his book to his old teacher Heinrich Wölfflin, whose enthusiasm was restrained.[184]

Despite the excellent critical reception, sales did not materialize for *Bauen in Frankreich*. During the first month after its publication some three hundred copies were sold, but by the end of the year this number had not even tripled. The publisher, Georg Biermann, brought up these difficulties in a letter to Giedion of 12 December 1928 and thought that they could be attributed to the choice of title:

> The sale of your book is very slow despite the Christmas advertising. They are a little despondent in Leipzig and cannot explain why a book that has received such stunning critical acclaim sells so miserably. I don't think 600 copies have been sold to date.[185] I am convinced that the title is to blame, and I seriously regret having allowed you to twist my arm at the last minute. I foresaw the consequences and can prove to you, in black and white, that in a case like this one should not reject in advance the wise experiences of an old publisher.[186]

Giedion replied on 29 December. He felt that the sales were depressing and insisted that the book was a "decent product." The many complimentary reviews proved that. He pleaded for an intensification of advertising ("we must proceed in the American fashion") and proposed the publication of a new prospectus (figs. v, vi). As for the title, he considered a change, which would be made in connection with the second edition. This would appear simultaneously with a French edition of the book and would be an expanded version of the original German text.[187] The title of the French edition would be *L'origine de l'architecture contemporaine* (The origin of contemporary architecture); the German title would correspondingly be *Ursprung des neuen Bauens*. Giedion's proposed change indicated that he suspected the tactical problem of connecting the "new architecture" with one country (France), which the book's German target audience would regard as less a prototype and more a competitor. Biermann agreed with this, apparently forgetting his old proposals for a title but not his reproaches against Giedion: "Believe me," he wrote Giedion on 5 January 1929, "the failure of your book can only be blamed on its foolish title. No German, not even a German architect, is interested in the new architecture in France, but yes, in the most important problem — building in iron and ferroconcrete. As long as this title is not changed, my feeling is that all efforts are in vain."[188] Biermann also insisted on this view in later letters sent to Giedion.

Giedion applied directly to Editions Crès for the first time on 3 April 1929. He referred to his correspondence with Le Corbusier who, as Giedion wrote, was the supervisor of a publication project for this publisher on the topic of "modern architecture," in which the translation of *Bauen in Frankreich* would be the first volume of a proposed series. For the French edition, Giedion held out the prospect of including additional unpublished documents. On 8 May 1929 Giedion contacted the architect Frank Otten in Brussels, whom Alberto Sartoris recommended to him for the French translation.[189] At the end of May Giedion traveled to France in order to discuss the plans for the French edition. On 29 June 1929 he wrote to Biermann to report that his discussions there had led to a proposal to publish *Bauen in Frankreich* with the new title *L'origine de l'architecture moderne.* The text would be expanded by twenty-four pages and the illustrations doubled. It would be the first volume of a new series, *Cycle d'architecture moderne.* The French publisher would apparently take care of a simultaneous German edition. Klinckhardt & Biermann would in that case be used only for sales in Germany.

These plans, however, were not realized. In the meantime, Giedion had embarked upon his next book project. Conceived as a multivolume work, *Die Entstehung des heutigen Menschen* (The origin of contemporary man), this history of modern civilization occupied him from the spring of 1929 until his departure for America—in 1938.

Notes

1. Reyner Banham, *Theory and Design in the First Machine Age* (London: Architectural Press, 1960; reprint, Cambridge: MIT Press, 1980), 309.

2. Spiro Kostof, "Architecture, You and Him: The Mark of Sigfried Giedion," *Daedalus* 105, no. 1 (1976): 189–204.

3. Stanislaus von Moos, "Dank an S. Giedion," *Neue Zürcher Zeitung*, 18 April 1968. Included in Paul Hofer and Ulrich Stucky, eds., *Hommage à Giedion: Profile seiner Persönlichkeit* (Basel: Birkhäuser, 1971), 144–52.

4. The systematic processing of Giedion's published and unpublished papers began with Stanislaus von Moos, who assembled a complete bibliography (see Hofer and Stucky [note 3], 187–98). This effort was given new impetus some years later, in 1979, when Giedion's entire literary estate was taken over by the Institut für Geschichte und Theorie der Architektur (GTA) at the Eidgenössische Technische Hochschule (ETH), Zurich.

Giedion's work has been widely reviewed. Major publications, in addition to *Hommage à Giedion*, include "Sigfried Giedion: Un progetto storico/Sigfried Giedion: A History Project," *Rassegna* 8, no. 25 (March 1986), with contributions by Stanislaus von Moos, Vittorio Magnago Lampugnani, Sokratis Georgiadis, Dorothee Huber, Gottfried Korff, and Joseph Rykwert. Sigfried Giedion, *Wege in die Öffentlichkeit: Aufsätze und unveröffentlichte Schriften aus den Jahren 1926–1956*, ed. Dorothee Huber (Zurich: Ammann, 1987). Sokratis Georgiadis, *Sigfried Giedion: Eine intellektuelle Biographie* (Zurich: Ammann, 1989); translated by Colin Hall under the title *Sigfried Giedion: An Intellectual Biography* (Edinburgh: Edinburgh Univ. Press, 1994). In 1989 the GTA organized a series of events at the ETH for the centennial celebration of Giedion's birth. A scientific colloquium was accompanied by two exhibitions. An exhibition catalog was also published: Institut für Geschichte und Theorie der Architektur, *Sigfried Giedion, 1888–1968: Der Entwurf einer modernen Tradition*, exh. cat. (Zurich: Ammann, 1989), with contributions by Jos Bosman, Sokratis Georgiadis, Dorothee Huber, Claude Lichtenstein, Friederike Mehlau-Wiebking, Werner Oechslin, Arthur Rüegg, and Joseph Rykwert.

5. Sigfried Giedion, *Spätbarocker und romantischer Klassizismus* (Munich: F. Bruckmann, 1922).

6. See also Friederike Mehlau-Wiebking, Arthur Rüegg, and Ruggero Tropeano, *Schweizer Typenmöbel 1925–1935: Sigfried Giedion und die Wohnbedarf AG* (Zurich: GTA, 1989).

7. For a detailed history of the construction of these houses, see Joachim Driller, "Mar-

cel Breuer: Das architektonische Frühwerk bis 1950" (Ph.D. diss., University of Freiburg, 1990).

8. See also Eduard F. Sekler, "Sigfried Giedion at Harvard University," in *The Architectural Historian in America*, Studies in the History of Art, ed. Elisabeth Blair MacDougall, vol. 35 (Washington, D.C.: National Gallery of Art, 1990), 265–73.

9. Sigfried Giedion, *Space, Time, and Architecture: The Growth of a New Tradition* (Cambridge: Harvard Univ. Press, 1941).

10. Kenneth Frampton, "Giedion in America: Reflections in a Mirror," in Demetri Porphyrios, ed., *On the Methodology of Architectural History* (London: Architectural Design, 1981), 44–51.

11. Sigfried Giedion, *Mechanization Takes Command: A Contribution to Anonymous History* (New York: Oxford Univ. Press, 1948).

12. See also Stanislaus von Moos, "Die zweite Entdeckung Amerikas: Zur Vorgeschichte von *Mechanization Takes Command*," Epilogue to Sigfried Giedion, *Die Herrschaft der Mechanisierung* (Frankfurt am Main: Europäische Verlagsanstalt, 1982), 779–816. See also Sokratis Georgiadis, "Sigfried Giedion: Patents in Historical Investigation," *Rassegna* 13, no. 46 (June 1991): 54–61.

13. Important texts on this topic are collected in Sigfried Giedion, *Architecture, You and Me: The Diary of a Development* (Cambridge: Harvard Univ. Press, 1958).

14. See also Sokratis Georgiadis, "Giedion: Il simbolo e il corpo," *Casabella* 55, no. 599 (March 1993): 48–51.

15. The fruits of these investigations are the books by Sigfried Giedion, *The Eternal Present: A Contribution on Constancy and Change* (New York: Bollingen Foundation, 1962), vol. 1, *The Beginnings of Art;* (New York: Bollingen Foundation, 1964), vol. 2, *The Beginnings of Architecture;* idem, *Architecture and the Phenomena of Transition* (Cambridge: Harvard Univ. Press, 1971).

16. Gottfried Semper, *Wissenschaft, Industrie und Kunst: Vorschläge zur Anregung nationalen Kunstgefühles, bei dem Schlusse der Londoner Industrie-Ausstellung, London, den 11. Oktober 1851* (Braunschweig: F. Vieweg & Sohn, 1852). Quoted from Gottfried Semper, *Wissenschaft, Industrie und Kunst, und andere Schriften über Architektur, Kunsthandwerk und Kunstunterricht*, Neue Bauhausbücher, ed. Hans Maria Wingler (Mainz: Kupferberg, 1966), 68.

17. Carl Gottlieb Wilhelm Bötticher, *Die Tektonik der Hellenen*, 2 vols. (Potsdam: Ferdinand Riegel, 1844).

18. Carl Gottlieb Wilhelm Bötticher, "Das Prinzip der Hellenischen und Germanischen Bauweise hinsichtlich der Übertragung in die Bauweise unserer Tage," in Julius Posener, ed., *Festreden: Schinkel zu Ehren, 1846–1980* (Berlin: Fröhlich & Kaufmann, [1981?]), 11ff.

19. Bötticher (see note 17), vol. 1, first excursus, "Über die Entwikkelung der freien Glieder des Baues und deren Einfluß auf die Bewältigung des Materials" (1843), 1: "... *den*

Verdacht von mir zu entfernen als habe mich eine unfreie und einseitige Neigung für die Gebilde der hellenischen Tektonik bei der Arbeit über ihre Prinzipien geleitet, und mir Auge und freien Blick genommen für den Werth der Tektonik des Mittelalters."

20. Wolfgang Herrmann has given a very informative survey of this debate in his Introduction to *In What Style Should We Build? The German Debate on Architectural Style* (Santa Monica: The Getty Center for the History of Art and the Humanities, 1992).

21. Bötticher (see note 18), 15: "*Es bezeichnen beide nur zwei Entwicklungsstufen, die erst vorangehen und ihren vorgezeichneten Kreis erfüllen mussten, bevor eine dritte Weise an das Licht treten kann, welche keine der vorigen negiert, sondern vielmehr nur auf beider Resultate sich gründen könne, um eine dritte und höhere Stufe der Entwicklung einzunehmen, als irgend eine von jenen erstiegen hatte; eine dritte Weise zu deren Erzeugung die uns folgende Zeit schon der geschichtlichen Notwendigkeit nach berufen ist und zu deren Beginn unsere Zeit in der Tat auch schon angehoben hat, den Grund zu legen.*"

22. Eduard Metzger, "Beitrag zur Zeitfrage: In welchem Stil man bauen soll!" *Allgemeine Bauzeitung* 10 (1845): 169–79, esp. 176: "*Eisenkonstruktzionen. Dieses Wort ist für den Plastiker als Architekten, ich glaub es gerne, ein Schreckenswort!*" (emphasis in original).

23. Ibid., 177: "*Das Netz des spitzbogigen Bausystems in seiner gesammten Verspannung ist jenem, das sich der Natur des Eisens gemäss entwickelt, nahe verwandt.*"

24. Metzger (see note 22), 177–78: "*Es scheint fast als ob Eisenverspannung und massenhafter Baukörper ausgesprochene Gegensätze sind.*"

25. Richard Streiter, *Karl Böttichers "Tektonik der Hellenen" als ästhetische und kunstgeschichtliche Theorie: Eine Kritik*, Beiträge zur Ästhetik, vol. 3 (Hamburg: Leopold Voss, 1896).

26. Richard Streiter, *Architektonische Zeitfragen: Eine Sammlung und Sichtung verschiedener Anschauungen mit besonderer Beziehung auf Professor Otto Wagners Schrift "Moderne Architektur"* (Berlin: Cosmos, 1898). Reprinted in idem, *Ausgewählte Schriften zur Aesthetik und Kunst-Geschichte*, ed. Franz von Reber and Emil Sulger-Gebing (Munich: Delphin, 1913), 55–149.

27. Heinrich Leibnitz, *Das struktive Element in der Architektur und sein Verhältnis zur Kunstform* (Tübingen: L. F. Fues, 1849), 80: "*Das Eisen ist seinem Stoff nach kein Material, das in der Architektur eine neue Kunstepoche hervorrufen könnte, denn es überschreitet das Maass, widerspricht dem statischen Gefühl und zerstört die Masse.*"

28. Gottfried Semper, "Ueber Baustile" (1869), in idem, *Kleine Schriften*, ed. Hans and Manfred Semper (Berlin: W. Spemann, 1884; reprint, Mittenwald: Mäander, 1979), 395–426.

29. Wolfgang Herrmann takes a critical look at this in "Stellung Sempers zum Baustoff Eisen," in idem, *Gottfried Semper: Theoretischer Nachlaß an der ETH Zürich: Katalog und Kommentare*, Geschichte und Theorie der Architektur, vol. 15 (Basel: Birkhäuser, 1981), 61–68.

30. Gottfried Semper, "Ueber Wintergärten" (1849), in idem, *Kleine Schriften* (see note 28), 484–90.

31. Gottfried Semper, *Der Stil in den technischen und tektonischen Künsten; oder, Praktische Ästhetik*, 2 vols. (Frankfurt am Main: Verlag für Kunst und Wissenschaft, 1860–1863), 2: 263–64: "*... mageren Boden für die Kunst! Von einem eigenen monumentalen Stab- und Gussmetallstil kann nicht die Rede sein; das Ideal desselben ist unsichtbare Architektur! Denn je dünner das Metall gespinnt, desto vollkommener in seiner Art*" (emphasis in original).

32. Semper (see note 31), 2: 550: "*Die gefährliche Idee, aus der Eisenkonstruktion, angewandt auf Monumentalbau, müsse für uns ein neuer Baustil hervorgehen, hat schon manchen talentvollen, aber der hohen Kunst entfremdeten Architekten auf Abwege geführt.*"

33. Semper criticized Labrouste's Bibliothèque Sainte-Geneviève in "Der Wintergarten zu Paris," in A. Romberg's *Zeitschrift für praktische Baukunst* (1849); partially reprinted under the title "Eisenkonstruktionen," in Wingler (see note 16), 22–24.

34. Semper (see note 31), 2: 551: "*Gestatten und loben wir den sichtbaren einfachen eisernen Dachstuhl der Eisenbahn-Ingenieurs bei Einsteighallen und sonstigen Schoppen, als Wahrzeichen ihres Provisoriums. Ersparen wir uns die Bewunderung gezierter eiserner Bibliotheken, Festsäle u. dergl.!*"

35. Ludwig Bohnstedt, 1822–1885.

36. Ludwig Bohnstedt, "Über die Bedeutung des Eisens für die Baukunst," *Deutsche Bauzeitung* 1 (1867): 201–2, 209–10, 219–20, esp. 220: "*Die Frage über die Schaffung eines neuen Baustiles, der dem Metalle seine Entstehung zu verdanken hätte, [dürfte] als erledigt zu betrachten sein; ein neuer Baustil in Folge der Verwendung des Eisens steht nicht zu erwarten.*"

37. Ibid.: "*Unsere alten üblichen Stilgesetze wurzeln eben in den Erfahrungen, welche wir an dem Vollmateriale, dem Steine, gesammelt und mit ihm in Einklang gebracht haben; sie bedingen die Erfüllung aller der Ansprüche, welchen bisher nur allein der Stein zu genügen vermochte*" (emphasis in original).

38. Cf. Dieter Dolgner, *Architektur im 19. Jahrhundert: Ludwig Bohnstedt — Leben und Werk* (Weimar: Hermann Böhlaus Nachfolger, 1979), 138f.

39. Richard Lucae, "Ueber die ästhetische Ausbildung der Eisen-Konstruktionen, besonders in ihrer Anwendung bei Räumen von bedeutender Spannweite," *Deutsche Bauzeitung* 4 (1870): 9–12.

40. Richard Lucae, "Über die Bedeutung und Macht des Raumes in der Baukunst," *Zeitschrift für praktische Baukunst* 29 (1869): 197ff.

41. Ibid., 199: "*. . . wird der Styl die Raumwirkung nur in einem sehr geringen Masse beeinflussen.*"

42. Lucae (see note 40), 202: "*Wäre den grossartigen Konstruktionsgedanken ihrer Deckenform zugleich ein bedeutungsvoller Schönheitsgedanke beigefügt worden, so würde unser Auge, das sich in dem sinnverwirrenden Durcheinander der sich überall durchkreuzenden eisernen Stäbe und*

eisernen Taue nicht zurecht finden kann, zur Ruhe gelangen und Genuss empfinden, wenn man, bildlich gesagt, unseren Blicken die einzelnen Exempel dieser in Eisen übersetzten Rechnung entzöge und nur das Resultat derselben in übersichtlicher Summe, zu einem System geordnet, in schöner Form zur Anschauung brächte, da ja die reine mathematische Konstruktion keine fertige Leistung der Kunst ist, sondern nur ein Gerippe, gleich dem des menschlichen Körpers."

43. Lucae (see note 39), 9: "*Die reine mathematische Konstruktion ist eben so wenig eine fertige Leistung der Kunst, als der menschliche Körper mit seinen offen liegenden Muskeln und Bändern, oder gar nur sein Gerippe ein lebensfähiges Geschöpf der Natur ist, und darum behaupte ich, die Schönheit der Bausysteme hat zum Theil darin ihren Grund, dass ein Ueberschuss an Masse über das zum Tragen nothwendige Material vorhanden ist*" (emphasis in original).

44. Lucae (see note 39), 9: "*Ein Geschlecht nach uns, welches so aufwächst mit der Eisen-Konstruktion, wie wir mit der Stein-Konstruktion aufgewachsen sind, wird in manchen Fällen das volle ungestörte Gefühl der Schönheit haben, in denen wir heute noch unbefriedigt bleiben, weil eine uns liebgewordene Schönheits-Tradition scheinbar angegriffen wird.*"

45. Lucae (see note 39), 12: "*Aber die höhere Aufgabe erblickt die Tektonik in den grossen allgemeinen Gedanken, die mit jenen einzelnen kleinen Symbolen—gewissermassen ihren Worten—ausgesprochen werden müssen!*"

46. Lucae (see note 39), 12: "*So paradox es im ersten Augenblicke klingen mag, man kann die Decke nur zeigen, wenn man sie mehr oder weniger versteckt.*"

47. Bötticher (see note 17), xivf.: "*. . . die werkthätige Hand des Tektonen jedes Glied in einem körperlichen Schema* [*erbildet*], *wie es sich zur Bildung der Räumlichkeit . . . entledige.*"

48. Constantin Lipsius, "Über die ästhetische Behandlung des Eisens im Hochbau," *Deutsche Bauzeitung* 12 (1878): 363–66.

49. Ibid., 363: "*Ein echtes Kind unserer Zeit, das sich gewaltig hervor gethan, ist unter den technischen Wissenschaften die Ingenieur-Wissenschaft. Auf durchaus modernem, realen Boden stehend, geht sie auf das ausschliesslich Zweckmässige aus, und dieses in rücksichtsloser Konsequenz anstrebend und in nacktester unerbittlicher Wahrheit—alles Aesthetische anderen Bestrebungen überlassend—zur Erscheinung bringend. Je geringer der Aufwand an Stoff, je minimaler die Dimensionen bei Erreichung maximalster Leistungen, um so grösser der Triumph! Und da das rein technisch Zweckliche der Verklärung durch die Schönheit nicht bedarf, weil sein Eintritt in die Erscheinung nur der Ausdruck der Funktion, die es zu verrichten, die Form gewordene Funktion ist, so tragen solche rein zweckliche Gebilde in ihrer Konstruktion selbst die Erklärung für ihr Vorhandensein, ihre Nothwendigkeit; sie überzeugen und befriedigen darum oft bis zu einem gewissen Grade ästhetisch.*"

50. Hubert Stier, "Rückblick auf die Entwicklung der deutschen Architektur in den letzten 50 Jahren," *Deutsche Bauzeitung* 26 (1892): 441–44, 446–53, 459–64.

51. Ibid., 447: "*. . . eben erst und zumeist nur als Gusseisen zu Schmuckzwecken anstelle des*

Schmiedeisens, dessen Behandlung man verlernt hatte, verwendet zu werden; von seiner Wichtigkeit für konstruktive Zwecke waren damals kaum die Anfänge einer Erkenntniss vorhanden."

52. Stier (see note 50), 452: "*... aber vorzugsweise nur als konstruktives Hilfsmittel unter Verkleidungen und Umhüllungen aus anderem Material. Man hatte noch nicht den Muth, dasselbe offen zu zeigen. Man schrieb über seine ästhetische Behandlung, aber man versuchte sie nicht."*

53. "Ein Beitrag zur Frage der Verwendung des Eisens im Hochbau," *Deutsche Bauzeitung* 17 (1883), 166: "*... dass man mittels des Gusseisens, sei es zu neuen Stilformen, sei es zu zeit- und ortsgemässen Umbildungen gewisser Elemente überkommener Stilformen, wie beispielsweise der maurischen gelangen werde."*

54. See Otto Peters, "Zur ästhetischen Behandlung von Eisenkonstruktionen," *Deutsche Bauzeitung* 17 (1883): 353–54.

55. See Puhlmann, "Zur ästhetischen Behandlung von Eisenkonstruktionen," *Deutsche Bauzeitung* 17 (1883): 425–27.

56. Georg Heuser, "Keime eines neuen Baustils," *Deutsche Bauzeitung* 22 (1888): 529–32, 534–35; idem, "Darwinistisches über Kunst und Technik," *Allgemeine Bauzeitung* 55 (1890): 17–19, 25–27; idem, "Der Gefachstil, eine werdende Bauart," *Deutsche Bauzeitung* 24 (1890): 565–76; idem, "Der Gefachstil in Beziehung zu Geräthen und Maschinen," *Deutsche Bauzeitung* 27 (1893): 149–54.

57. Georg Heuser, "Ein Nachwort zu den 'Stilbetrachtungen,'" *Deutsche Bauzeitung* 24 (1890): 626–31, esp. 627: "*Je nach der Gestaltung des Steges wechselt nun der Ausdruck der Gefachstile wie folgt:*

1. *Besteht der Steg aus dem für den 'Eisenstil' so charakteristischen Gitter- und Stab-Gefüge, so kommt jene u n s i c h t b a r e A r c h i t e k t u r zur Erscheinung....*
2. *Ist der Steg aus einer dünnen, aber vorwiegend g e s c h l o s s e n e n W a n d hergestellt, so entsteht zwar eine k ö r p e r l o s e, s t o f f a r m e, j e d o c h auch beim Eisen sehr s i c h t b a r e A r c h i t e k t u r....*
3. *Besteht der Steg aus voller Masse in verschiedenen Rohstoffen, so haben wir eine k ö r p e r l i c h e, eine M o n u m e n t a l - A r c h i t e k t u r vor Augen, bei welcher das Prinzip 'Gurt und Steg' sowohl mit seinem Vorzug der S t a b i l i t ä t, wie auch mit seinem gegensätzlichen Zierreichthum als Kunstgedanke verwerthet werden kann"* (emphasis in original).

The essay is an answer to K. E. O. Fritsch, "Stil-Betrachtungen," *Deutsche Bauzeitung* 24 (1890): 417–31, 434–40. Heuser's theory is evidently influenced by Salli Philipp's thoughts on architectural history. According to Philipp, "*lassen sich in Betreff der ästhetischen Ausbildung sämmtliche vorhandenen, sowie sämmtliche möglichen Baustile nach zwei Prinzipien, sowie den Uebergängen von dem einen Prinzip in das andere unterscheiden.... Diese beiden Prinzipien heissen 'S t ü t z e u n d L a s t' einerseits und 'R a h m e n u n d F ü l l u n g' andrerseits*" (All existing as well as all possible building styles may, with respect to aesthetic formation, be divided according to two principles, and the transition from one principle to another.... These two principles are called "*support and load*," on the one hand, and "*frame and filling*," on the other hand); (emphasis in original). See Salli Philipp, "Das Roccoco und die allgemeinen Prinzipien der Baustile," *Deutsche Bauzeitung* 13 (1879): 278–80, 288–91, 298–301, 308–11, esp. 288.

58. Robert Neumann, *Architektonische Betrachtungen eines deutschen Baumeisters, mit besonderer Beziehung auf deutsches Wesen und deutsche Baukunst* (Berlin: W. Ernst & Sohn, 1896). On the subject of iron, see esp. pp. 298–302. Despite the often overheated and blatant nationalism of the remarks, the author tries throughout much of the book to be straightforward.

59. Ibid., 301: "*Da die Hauptverwendung des Eisens für Monumentalbauten wohl stets die zur Bildung der Raumdecke sein wird, diese aber die Leistungsfähigkeit des Eisens bei Weitem nicht voll in Anspruch nimmt, so wird man den Einfluss des Eisens auf die Stilbildung nicht überschätzen dürfen.*"

60. Neumann (see note 58), 299: "*Am Äusseren der Bauwerke wird das Eisen sich wenig bemerklich machen, daher auch auf die Aussenarchitektur von geringem Einflusse bleiben. Hier wird der Stein, und vor Allem der kräftig wirkende Haustein sein altes Recht behaupten.*"

61. Adolf Göller, *Die Entstehung der architektonischen Stilformen: Eine Geschichte der Baukunst nach dem Werden und Wandern der Formgedanken* (Stuttgart: Konrad Wittwer, 1888), esp. 450f.

62. Ibid., 450: "*Ein neuer Stil unserer Zeit dürfte zwar aus dieser Vereinigung entgegen manchen Hoffnungen nicht zu erwarten sein, einmal weil das Schmiedeisen an den meisten Gebäudegattungen für die Wandkonstruktion niemals viel leisten wird, zum anderen weil es in seiner geringen Masse niemals Form genug darbieten kann und also in einem solchen Stil der Stein nach wie vor fast ausschliesslich deren Träger sein müsste*" (emphasis in original).

63. Fritsch (see note 57).

64. Fritsch (see note 57), 429: "*Ich muss mich dem gegenüber leider als hartnäckiger Ketzer bekennen. Denn mir scheint ein derartiger Glaube nicht nur als eine masslose Überschätzung, sondern vor allem als eine vollständige Verkennung der Bedeutung, welche dem Eisen in der Baukunst zukommt.*"

65. Fritsch (see note 57), 428: "*Dass sich dagegen die vor 100 Jahren begonnenen Stil-Experimente vorzugsweise den beiden Gerüst-Stilen der hellenischen und der gothischen Kunst zuwendeten, findet seinen Grund einfach darin, dass es eine Eigenthümlichkeit dieser Stile im Gegensatz zu den Massen-Stilen ist, einen Kanon zu besitzen. Ergiebt sich derselbe doch dadurch, dass es bis zu einem gewissen Grade Bedingung ist, das architektonische Gerüst in die Erscheinung treten zu lassen, während der Massen-Stil dem individuellen Empfinden des Künstlers erheblich grössere Freiheit gestattet. Für schwache, eines Anhalts bedürftige Kräfte wird es daher stets bequemer sein, eines Gerüst-Stils sich zu bedienen und es war gleichsam eine Natur-Nothwendigkeit, dass ein Zeitalter, welches vor allem nach festen Grundsätzen des künstlerischen Schaffen[s] verlangte, begierig nach ihnen griff*" (emphasis in original).

66. Stier (see note 50), 464: "*Auch das Eisen ist uns nicht nur ein Gehilfe geworden, der uns namentlich die Schaffung weiter freier Räume in früher nie für möglich gehaltenen Abmessungen gestattet: auch mit seiner künstlerischen Durchbildung haben wir uns befasst. Allerdings muss ich es aussprechen, dass gerade auf diesem Gebiete, wo die beiden grossen Richtungen des Fachs, die Kunst und das Ingenieurwesen recht eigentlich vereint arbeiten müssten, bei uns in Deutschland eine solche gemeinsame Thätigkeit etwas vermisst wird.*"

67. Stier (see note 50), 464: "*Mag unsere jüngere Schwester, die Ingenieurwissenschaft, in kürzerer Frist grössere und in die Augen springendere Erfolge aufzuweisen haben: das wollen wir ihr freudig und neidlos zugestehen; sind doch auch die Vortheile gross, die uns daraus erwachsen. Dem Baukünstler aber haben die Jahrtausende der Kulturgeschichte der Menschheit seinen Adelsbrief geschrieben und wenn nach 50 Jahren wieder ein Redner an dieser Stelle rückschauend spricht, so soll er uns bezeugen, dass auch wir verstanden haben, ihn hochzuhalten!*"

68. Albert Hofmann, "Die künstlerischen Beziehungen der Architektur zur Ingenieur-Wissenschaft," *Deutsche Bauzeitung* 27 (1893): 284–87, 289–91, 296–99, 301–3, esp. 284: "*Ist es möglich, dass der Herkules Ingenieur an dem Rocken der sanften Schönheit Omphale sitzt, dass die freie, heitere Kunst der Architektur in Beziehung treten kann mit dem nüchternen, herben Ernst der Ingenieur-Wissenschaft?*"

69. Ibid., 286: "*Alles ist in Umbildung begriffen und diese Umbildung, die fortwährend vor sich geht, ist in gleichem Maasse eine wirthschaftliche, eine wissenschaftliche, eine politische und eine philosophische.*"

70. Robert Dohme, *Das englische Haus: Eine kultur- und baugeschichtliche Skizze* (Braunschweig: n.p., 1888).

71. Hofmann (see note 68), 287: "*Der Ingenieur dringt erobernd vorwärts und es ist nicht zu leugnen: die Welt ist gefangen durch die kühnen und schönen Werke des Ingenieurs.*"

72. Hofmann (see note 68), 287: "*Können eiserne Brücken schön sein?*"

73. Hofmann (see note 68), 290: "*Der architecte constructeur war immer die Seele, der architecte décorateur das Kleid für diese Seele.*"

74. Hofmann (see note 68), 290: "*Heute ist in vielen Fällen das Kleid zur Hauptsache geworden, die Seele hat ein anderer geraubt. . . . Der Architekt droht zu sinken, der Ingenieur steigt unverhältnismässig. Die scharfe, rechnerische, verstandesmässige Geistesthätigkeit, unterstützt durch das sich mehr und mehr entwickelnde Schönheitsgefühl bilden gegenüber der freien, künstlerischen Gefühlsthätigkeit des Architekten das hebende Moment.*"

75. Hofmann (see note 68), 291: "*. . . freilich vorausgesetzt, dass der Grundsatz nicht angegriffen wird, dass die Schönheit der Linie der höchste Schönheits-Begriff ist.*"

76. Hofmann (see note 68), 296: "*Die Kunst liess sich bisher mehr vom Gefühle leiten, warum sollte sie sich nicht die Fortschritte des Denkens, der exakten Wissenschaften zu eigen machen? Hier klaffen und vereinigen sich die Gegensätze, Fühlen und Denken klammern sich an dieselbe Erscheinung und versöhnen sich.*"

77. Hofmann (see note 68), 301: "*Aus dem* Gegensatz *zwischen beiden werden die Bedürfnisse eine* Annäherung *schaffen, die Rivalen von heute arbeiten morgen zusammen. Und aus der Zusammenarbeit des Geschmacks und der Mathematik, des Formal-Schönen und der Kühnheit der Technik entsteht eine neue Kunst, ein Bild unserer Kultur, ähnlich Euphorion, dem symbolischen Sohn von Faust und Helena, welcher in sich den Geist der alten und der neuen Zeit, die Begeisterung für das Schöne und den Sinn für das Nützliche vereinigt*" (emphasis in original).

78. Gustav Lang, "Wissenschaft und Wirklichkeit im Bauwesen," *Deutsche Bauzeitung* 25 (1891): 563–64.

79. Ibid., 563: "*. . . 'um das Ding an sich, was weder schön noch hässlich ist', sondern um Gebilde von Menschenhand, die den Naturgesetzen entsprechen müssen, in denen unser ganzes Fühlen und Denken wurzelt, so dass wir nichts für schön halten können, was nach diesen Gesetzen als zweckwidrig erscheint.*"

80. Lang (see note 78), 563–64: "*Ein Widerspruch zwischen den Erzeugnissen einer richtigen Berechnung und eines gesunden Schönheitsgefühls ist thatsächlich gar nicht vorhanden. . . . Falls unsere Rechnung zu unschönen Formen führt, deutet dies darauf hin, dass nicht alle inbetracht kommenden Einflüsse beim Ansatz unserer Formeln richtig abgewogen wurden; das Schönheitsgefühl kann uns daher ein Leitstern bleiben, der unsere Rechnung davor bewahrt, sich in Einseitigkeiten zu verlieren*" (the entire quotation is emphasized in the original, for there it is a quotation from another publication by Gustav Lang).

81. Cornelius Gurlitt, *Die deutsche Kunst des neunzehnten Jahrhunderts: Ihre Ziele und Thaten*, 2nd ed. (Berlin: Georg Bondi, 1900). An abridged version was published under the title *Zur Befreiung der Baukunst: Ziele und Taten deutscher Architekten im 19. Jahrhundert*, Bauwelt Fundamente, ed. Werner Kallmorgen, vol. 22 (Berlin: Bertelsmann Fachverlag, 1969), 148: "*Nicht wir haben künstlerisch die Werkform des Eisens besiegt, sondern diese hat uns besiegt und gezwungen, sie für schön zu nehmen, da sie verständig und das Werk eines schaffenden Gedankens ist.*"

82. Cornelius Gurlitt, "Karl Bötticher," *Deutsche Bauzeitung* 24 (1890): 384–87, 393–95, esp. 384: "*Der Hellenismus scheidet aus der deutschen Nation,*" and 386: "*Die Tage des Individualismus brechen wieder an!*"

83. Hermann Muthesius, *Stilarchitektur und Baukunst* (Mülheim a.d. Ruhr: K. Schimmelpfeng, 1902), 50–51: "*Will man daher nach einem neuen Stile, dem Stile unserer Zeit suchen, so wäre den Kennzeichen desselben viel eher in solchen neuartigen Schöpfungen nachzuspüren, die wirklich ganz neu entstandenen Bedürfnissen dienen, wie etwa in unsern Bahnhöfen, Ausstellungsbauten, Riesenversammlungshäusern, ferner auf allgemein-tektonischem Gebiete, in unsern Riesenbrücken, Dampfschiffen, Eisenbahnwagen, Fahrrädern usw. In der That sehen wir gerade hier wirklich neuzeitliche Gedanken und neue Gestaltungsgrundsätze verkörpert, die uns zu denken geben müssen. Wir bemerken eine strenge, man möchte sagen, wissenschaftliche Sach-*

lichkeit, eine Enthaltung von allen äussern Schmuckformen, eine Gestaltung genau nach dem Zweck, dem das Werk dienen soll. Und trotzdem, wer möchte den gefälligen Eindruck einer weit geschwungenen Eisenbrücke leugnen, wem gefällt nicht der heutige elegante Landauer, das schmucke Kriegsschiff, das zierliche Zweirad?" Translated and edited by Stanford Anderson under the title *Style-Architecture and Building-Art* (Santa Monica: The Getty Center for the History of Art and the Humanities, 1995), 79.

84. Hermann Muthesius, "Das Formproblem im Ingenieurbau," *Jahrbuch des Deutschen Werkbundes*, 1913: 23–32; quoted from Wend Fischer, ed., *Zwischen Kunst und Industrie: Der Deutsche Werkbund* (1975; reprint, Stuttgart: Deutsche Verlags-Anstalt, 1987), 74ff.

85. Ibid., 77: "*Selbst den Fall angenommen, dass lediglich der Gebrauchszweck vorgeschwebt hätte, so lässt sich doch behaupten, dass den Verfertiger, sei es auch nur aus einem von ihm selbst nicht gefühlten ästhetischen 'Unterbewusstsein' heraus, auch die Rücksichten auf die Form beeinflusst haben.*"

86. Muthesius (see note 84), 76: "*Die Vorstellung, es genüge für den Ingenieur völlig, dass ein Bauwerk, ein Gerät, eine Maschine, die er schafft, einen Zweck erfülle, ist irrig, noch irriger ist der neuerdings oft gehörte Satz, dass wenn sie einen Zweck erfülle, sie zugleich auch schön sei. Nützlichkeit hat an und für sich nichts mit Schönheit zu tun.*"

87. Albert Hofmann, "Kunst- und Ingenieurwesen," *Deutsche Bauzeitung* 42 (1908): 98–100 (speech given to the Architekten-Verein zu Berlin 16 December 1907), esp. 98–99: "*Vom Ingenieurwesen erwarten wir die Bereicherung der Kunst durch neue Formen, neue Bildungen als äusseren Ausdruck einer neuen, im Inneren der Geister und ihrer Werke wirkenden Macht.*" See also *Wochenschrift des Architekten-Vereins zu Berlin* 3 (1908): 61ff., 68ff.

88. Friedrich Naumann, "Die Kunst im Zeitalter der Maschine," *Schweizerische Bauzeitung* 44 (1904): 112ff., 123ff., esp. 116: "*Hier wird nicht Kunst neben der Konstruktion getrieben, keine angeklebte Dekoration, keine blosse Schnörkelei, hier wird für den Zweck geschaffen, und die Form wird geboren wie ein Kind, an das seine Eltern kaum dachten.*" First published in *Kunstwart* 17, no. 20 (1904).

89. Ibid., 16: "*Hier leben noch unaussprechliche Möglichkeiten. Alle alten Raumbegriffe verschieben sich. Träger und Belastungsverhältnisse werden anders. Grosse Gewölbe fast auf Punkte zu legen, ist so neu, dass oft der Architekt noch falsche Pfeiler für nötig hält, als schäme er sich selbst seiner jungen Kraft. Noch gibt man dem Eisenbau aus einer Art von Schüchternheit steinerne Vorhallen. Gerade aber dieses leise und doch so frohe Herauskommen aus dem Walde der Vergangenheit gehört mit zum Zauber der neuen Kunst.*"

90. Heinrich Pudor, "Zur Ästhetik der Eisenarchitektur," *Der Architekt* 8 (1902): 1–3; idem, "Gedanken über moderne Architektur," *Der Architekt* 8 (1902): 13–15; idem, "Erziehung zur Eisenarchitektur," *Der Architekt* 9 (1903): 24–26.

91. Heinrich Pudor, "Die Schönheit der Maschinen und Eisenbauten," *Zeitschrift für Architektur und Ingenieurwesen* 56, no. 2, n.s., 15 (1910): cols. 141–50, esp. 143: "*Der Material-*

stil, der uns von dem überflüssigen Dekor und Ornament befreit und die schöne Zweckform mit dem schönen Material verbindet, bildet die moderne Ausdrucksform nicht nur für das Kunstgewerbe, sondern für die Gewerbekunst und Industrie, und gleichfalls für die Architektur und Ingenieurkunst."

92. Ibid., col. 145: "*Auch hier ist die erste Periode dadurch gekennzeichnet, dass man sich der Eisenkonstruktion schämte und sie den Blicken zu verbergen suchte, während man sie in der zweiten Periode, deren Anfang durch den Eiffelturm gekennzeichnet ist, sie den Augen in ihrer nackten Schöne blosslegte.*"

93. Wilhelm Freiherr von Tettau, "Ästhetik des Eisens," *Deutsche Bauhütte* 8, no. 28 (1904): 190f., esp. 190: "*Sollte mit dem technisch vollendeten auch der künstlerisch vollendete Ausdruck identisch sein?*" See also idem, "Zur Aesthetik der Eisenarchitektur," *Deutsche Bauzeitung* 42, no. 4 (1908): 24–26.

94. Tettau, "Zur Aesthetik der Eisenarchitektur" (see note 93), 26: "*... nach meiner Überzeugung [steigert sich] der ästhetische Genuss mit der grösseren Verständlichkeit des Systems.*"

95. Streiter (see note 26).

96. Otto Wagner, *Modern Architecture*, ed. and trans. Harry Francis Mallgrave (Santa Monica: The Getty Center for the History of Art and the Humanities, 1988), 98.

97. Streiter (see note 26), 109–10: "*... dass nicht die neue Technik einen neuen Stil hervorbringt, sondern dass ein bestimmtes Formgefühl die neue Technik nach dieser oder jener Richtung künstlerisch ausbildet.*"

98. Streiter (see note 26), 110: "*Wo immer Eisenkonstruktionen in bedeutenden Abmessungen offen liegend und für sich allein auftreten, da zeigt sich ihre absolute Sprödigkeit gegen künstlerische Gestaltung. Die Hoffnung, dass die Zukunft das bis zur Stunde nicht Erreichte bringen wird, kann mit Sicherheit als trügerisch bezeichnet werden. Denn die Möglichkeit, eine Eisenkonstruktion durch struktur-symbolische Formung des Ganzen, wie namentlich der Teile zu einem nicht mehr abstrakt-starren, sondern konkret-lebendigen statischen Organismus auszubilden, dessen Glieder durch den Ausdruck ihrer Körperlichkeit uns ihr Zusammenwirken zu einem Zustand 'glücklichen Gleichgewichts' unmittelbar fühlen lassen, diese einzige Möglichkeit tektonischer 'Idealisierung' der Werkform ist durch die Natur des Materials und durch die Art seiner Zusammenfügung ausgeschlossen. Die fleischlose Dünnheit und steife Trockenheit der Konstruktionsteile, die durch die statische Berechnung gegebene Gebundenheit der Anordnung, die äussere Gleichartigkeit der Glieder, welche die Verschiedenheit ihrer Leistung (Widerstand gegen Zug oder Druck) im allgemeinen nicht erkennen lassen, die bei grossen Konstruktionen verwirrende Menge sich durchkreuzender fast körperloser Linien, deren Sinn und Zweck nur dem technisch geschulten Verstand nicht aber dem einfachen Gefühl fassbar wird: all das lässt uns die Eisenkonstruktionen gleichgültig erscheinen, wenn auch manchmal der Gesamtumrisslinie solcher Werke (Bogenbrücken, Eiffel-Turm) oder der Wirkung der durch sie ermöglichten kolossalen Innenräume ein gewisser ästhetischer Reiz nicht abzusprechen ist. Eine bedeutende, tiefgehende, echt künstlerische*

Stimmung hervorzurufen wird aber auch der grossartigsten Eisenkonstruktion nicht gelingen."

99. "Aufruf zur Gründung eines Bundes Heimatschutz," *Deutsche Bauhütte* 8, no. 15 (1904): 106f.: "*Den Zauber einsamer Gebirgswelt vernichtet man durch aufdringliche Bauten. Eiserne Brücken spannt man in unschönen, das Landschaftsbild verunstaltenden Formen über unsere Wasserläufe, auch da, wo allen Anforderungen der Zweckmässigkeit mit schlichten Stein- und Holzbrücken zu entsprechen gewesen wäre.*"

100. See Hartmut Frank, "Heimatschutz und typologisches Entwerfen," in Vittorio Magnago Lampugnani and Romana Schneider, eds., *Moderne Architektur in Deutschland 1900 bis 1950*, exh. cat. (Stuttgart: Gerd Hatje, 1992), vol. 1, *Reform und Tradition*, 105–31.

101. Joseph August Lux, *Ingenieur-Aesthetik* (Munich: G. Lammers, 1910).

102. Filippo Tommaso Marinetti's manifesto, "Le futurisme," was published in *Le Figaro*, 20 February 1909.

103. Lux (see note 101), 1 and 35: "*Wir können nichts dafür, dass uns heute schon von Kindesbeinen an die Technik wichtiger ist als Plato. . . . Zwei Welten stehen einander schroff gegenüber.*"

104. Lux (see note 101), 4: "*Soll nun die Technik, die durchaus Kind unserer Zeit ist, die alten Moden tragen, die alten, längst vergangenen Stile für ihren jugendlichen riesenhaften Körper zurechtschneidern und sich tragen wie die Urgrossmutter? Es ist ein gesunder Instinkt, der diese Rückwärtsbewegung nach alten Motiven gründlich ablehnt. Die künstlerische Form muss aus den neuen Elementen neu erfunden werden. Das ist das Problem, an dem wir alle arbeiten*" (emphasis in original).

105. Lux (see note 101), 53: "*Nicht in der Architektur, sondern in den Fahrzeugen, in der modernen Verkehrstechnik spiegelt sich unsere Kultur. Wenn wir nach dem Stil unserer Zeit fragen, hier haben wir ihn.*"

106. Lux (see note 101), 14: "*Der eigentliche Architekt der modernen Zeit ist der Ingenieur.*" For (p. 45): "*nicht die Architekten, sondern die Ingenieure haben die Eisensprache entdeckt.*"

107. Lux (see note 101), 14: "*Dabei aber darf nicht die irrige Meinung Platz greifen, dass das technische Genie ohne Rücksicht auf die ästhetische oder, wenn wir wollen, künstlerische Erscheinung arbeitet. Es ist sehr vieles an Konstruktionen und Maschinen, das seine Form nicht der blossen mathematischen Rechnung, sondern einer Empirie verdankt, die die Gestaltung gewisser Teile dem Formgefühl überlässt.*"

108. Alfred Gotthold Meyer, *Eisenbauten: Ihre Geschichte und Aesthetik* (Esslingen: Paul Neff, 1907). In his book *Pioneers of Modern Design: From William Morris to Walter Gropius* (Harmondsworth: Penguin, 1982), 118 n. 2, Nikolaus Pevsner was the first to claim a dependence of Sigfried Giedion's *Bauen in Frankreich* on Meyer's *Eisenbauten*, but he was at a loss for proof. The same relationship was the central theme of Jos Bosman's paper "Der Ingenieur, der 'Stütze und Last' bekämpft," in *Sigfried Giedion, 1888–1968: Der Entwurf einer modernen Tradition* (see note 4), 55–70.

109. Meyer (see note 108), 43: "*Das Wesen des Eisenbaues beruht ja auf der 'rationellen Konstruktion', durch die er der Beanspruchung mit weitaus geringerem Materialvolumen zu genügen vermag, als alle anderen Baustoffe.*"

110. Meyer (see note 108), 43: "*Sie steht... gewissermassen nur am Ende jenes synthetischen Weges, der die Probleme der Mechanik aus dem Bereich arithmetischer Operationen und algebraischer Formeln in die Anschauung graphischer Gebilde überträgt.*"

111. Meyer (see note 108), 47: "*Solches 'Rechnen' wird dann ein Disponieren von Kräften, bei dem* nicht *mehr das zahlenmässig und technisch 'Rationelle' gesucht wird, sondern das formal günstig* Wirkende" (emphasis in original).

112. August Schmarsow, "The Essence of Architectural Creation," in *Empathy, Form, and Space*, ed. and trans. Harry Francis Mallgrave and Eleftherios Ikonomou (Santa Monica: The Getty Center for the History of Art and the Humanities, 1994), 281–97. Originally published as *Das Wesen der architektonischen Schöpfung* (Leipzig: Karl W. Hiersemann, 1894).

113. Meyer (see note 108), 49: "*Die 'Baukunst' also, die auf diesem Wege entsteht, wird nie und nimmer die monumentale Stilarchitektur etwa verdrängen und ersetzen können, wie einige allzu feurige Propheten des Eisenbaues verkünden.*"

114. Karl Scheffler, *Moderne Baukunst* (Berlin: Julius Bard, 1907). See esp. the first chapter, "Stein und Eisen."

115. Ibid., 10: "*Konstruktion ist nicht Kunst.*"

116. Scheffler (see note 114), 10: "*Der Künstler entmaterialisiert die Materie.*"

117. Scheffler (see note 114), 9: "*... dass es nie eine Monumentalarchitektur gegeben hat, die sich nicht des Steins bedient hätte.*"

118. Scheffler (see note 114), 15: "*... und noch mehr schliesst es darum die freie Kunstform aus.*"

119. Scheffler (see note 114), 19: "*Die Linie bedeutet in der Baukunst nichts, die Masse alles.*"

120. Scheffler (see note 114), 16: "*Dem Eisenbau ist eine Art von Schönheit zweiten Grades eigen, eine gebundene Konstruktionsschönheit und eine besondere Stilidee. Die Konsequenz: das ist der Stil des Eisens.*"

121. Scheffler (see note 114), 17: "*Es ist Wahnsinn, die höhere Ästhetik retten zu wollen, indem man den Eisenkonstruktionen historische Kunstformen gesellt.*"

122. Scheffler (see note 114), 22: "*Wenn wir einst eine eigene grosse Baukunst haben sollten, wird ihre [der Eisenbauten] Eigenart sicherlich in mehr als einem Zug an den herben, fast gotischen Ernst der Ingenieure erinnern. Es wird noch eine weite Wanderung nötig sein, bevor dauernde Resultate erzielt werden können; das Eisen wird der Baukunst der Zukunft aber um so bessere Dienste leisten, je weniger es beansprucht, ein Kunstmaterial zu sein.*"

123. Hermann Jordan, *Die künstlerische Gestaltung von Eisenkonstruktionen*, 2 vols. (Berlin: C. Heymann, 1913).

124. Ibid., 67–71.

125. Hermann Muthesius, *Kultur und Kunst: Gesammelte Aufsätze über künstlerische Fragen der Gegenwart* (Jena: E. Diederichs, 1904).

126. Jordan (see note 123), 1: 68: "*Wissenschaftliche Erwägungen und Berechnungen können dabei oft wenig helfen.*"

127. Jordan bases this on Karl Groos's *Einleitung in die Ästhetik* (Giessen: J. Ricker, 1892).

128. Jordan (see note 123), 1: 70: "*Um diese Wirkung näher zu erläutern, muss ich wiederum auf das schon mehrfach erwähnte Werk von Groos zurückgreifen. Unter den verschiedenen ästhetischen Modifikationen erläutert Groos auch den Begriff des Erhabenen. Er kommt bei seinen Untersuchungen zu dem Schlusse, dass zur Eigenschaft des Erhabenen einerseits gewaltige, über das Durchschnittmass hinausgehende Abmessungen, andererseits einfache Formen erforderlich sind.*

'Das Erhabene ist ein Gewaltiges in einfacher Form'" (emphasis in original).

129. E. von Mecenseffy, ed., *Die künstlerische Gestaltung der Eisenbetonbauten*, supplementary vol. 1 of *Handbuch für Eisenbetonbau*, ed. Fritz von Emperger (Berlin: W. Ernst & Sohn, 1911).

130. Ibid.: "*Dieses Zusammentreffen scheint Ursache gewesen zu sein, dass man dem Eisenbeton unbefangener entgegenkam als z.B. seinem unmittelbaren Vorgänger unter den neuen Baustoffen, dem Eisen, nach dessen angemessener Gestaltung ja noch heutigentags mit heissem Bemühen gerungen wird.*"

131. Mecenseffy (see note 129): "*Noch heute empfinden wir solche Gebilde als hässlich und fremd.*"

132. See Peter Collins, *Concrete: The Vision of a New Architecture: A Study of Auguste Perret and His Precursors* (London: Faber & Faber, 1959), 124–26. See also Alexander von Wielemans, "Der Eisenbetonbau in der Monumentalarchitektur," *Zeitschrift des österreichischen Ingenieur- und Architekten-Vereins* 61 (1909): 281–86. And idem, "Der Betoneisenbau in der Monumentalarchitektur," *Architektonische Rundschau* 25 (1909): 49–52, 59–61.

133. Collins (see note 132), 128.

134. These opinions are noted in Deutscher Werkbund, *Die Durchgeistigung der deutschen Arbeit: Ein Bericht* (Jena: E. Diederichs, 1911).

135. Werner Oechslin was the first to point out this relationship: "Allemagne — Influences, confluences et reniements," in Jacques Lucan, ed., *Le Corbusier, une encyclopédie*, exh. cat. (Paris: Centre Georges Pompidou, 1987), 33–39.

136. Peter Meyer, "Schönheit und Konstruktion," *Schweizerische Bauzeitung* 92, no. 17 (18 August 1928): 83–86, 91.

137. Walter Curt Behrendt, *Der Sieg des neuen Baustils* (Stuttgart: Dr. F. Wedekind & Co., 1927), 5: "*... Gebilde von einfach strenger Form und übersichtlichem Aufbau, mit glatten flächigen Mauern, mit durchwegs flachem Dach und geraden Umrisslinien. Die Gliederung der Baukör-*

per wird in der Regel nur durch eine mehr oder weniger bewegte Stufung der Baumassen bewirkt, durch Verteilung der Fenster und Öffnungen auf der Mauerfläche."

138. Ibid. That Behrendt's interpretation was not sheer invention, but rather corresponded completely to what the architects had in mind, is evident from Le Corbusier's *Vers une architecture.* Le Corbusier opens his "Three Reminders to Architects" with remarks on mass. Accompanied by illustrations of massive grain silos is his famous definition of architecture as "the masterly, correct and magnificent play of masses brought together in light." See Le Corbusier, *Towards a New Architecture,* trans. Frederick Etchells (London: John Rodker, 1931; reprint, New York: Dover, 1986), 29.

139. Julius Vischer and Ludwig Hilberseimer, *Beton als Gestalter: Bauten in Eisenbeton und ihre architektonische Gestaltung: Ausgeführte Eisenbetonbauten* (Stuttgart: J. Hoffmann, 1928). The same remark also holds for the first monograph on Perret's work (cited by Giedion), which was published the previous year: Paul Jamot, *A. G. Perret et l'architecture du beton armé* (Paris: G. Vanoest, 1927).

140. It is not mere coincidence that *Bauen in Frankreich* resulted in Giedion's longstanding and intensive commitment to architectural photography.

141. Sigfried Giedion, *Bauen in Frankreich, Bauen in Eisen, Bauen in Eisenbeton* (Leipzig: Klinkhardt & Biermann, 1928), caption to figure 1 (see below, p. 90): "[*Dieser Bau*] *kann nicht aus dem Stadtbild fortgeleugnet werden, dessen phantastische Krönung er bedeutet. Aber sein Zusammenwirken mit der Stadt ist weder 'räumlich' noch 'plastisch' fassbar. Es entstehen schwebende Beziehungen und Durchdringungen. Die Grenzen der Architektur verwischen sich"* (emphasis in original).

142. Giedion (see note 141), 85 (below, p. 169): "*Die Häuser Corbusiers sind weder räumlich noch plastisch: Luft weht durch sie! Luft wird konstituierender Faktor! Es gilt dafür weder Raum noch Plastik, nur BEZIEHUNG und DURCHDRINGUNG! Es gibt nur einen einzigen unteilbaren Raum. Zwischen Innen und Aussen fallen die Schalen*" (emphasis in original).

143. Behrendt (see note 137), 42: "*Es bleibt also kein anderer Weg, als von neuem zu beginnen, als die alten, ungültig gewordenen Formbegriffe aufzugeben und in eigener selbstständiger Weise konstruierend, bildend, gestaltend zu Werke zu gehen*" (emphasis in original).

144. Giedion (see note 141), 6 (below, p. 90): "*Es scheint uns fraglich, ob der beschränkte Begriff 'Architektur' überhaupt bestehen bleiben wird*" (emphasis in original).

145. Reconstructing the publication history of *Bauen in Frankreich* was made possible by Giedion's correspondence, which is housed in the Archiv S. Giedion, Institut für Geschichte und Theorie der Architektur at the ETH-Hönggerberg, Zurich (henceforth abbreviated "Archiv S. Giedion"). All letters cited in this section are from that archive.

146. Archiv S. Giedion, letter from Giedion, 6 October 1926: "*An Architekturaufsätzen von meiner Seite soll es Ihnen nicht fehlen.*"

147. Ibid.: "*Ich arbeite an meinem Buch über die neue Architektur, das vielleicht einigermassen grundlegend wird, wenn ich die Intensität habe, es in allen Teilen gleichmässig durchzubluten.*"

148. Sigfried Giedion, "Zur Situation der französischen Architektur," *Cicerone* 19 (1927): 15–23, 174–89, 310–17.

149. Bruno Taut, *Bauen: Der neue Wohnbau* (Leipzig: Klinkhardt & Biermann, 1927).

150. Archiv S. Giedion, letter from Georg Biermann, 16 March 1927: "*Auf die Buch-Idee komme ich zurück, nachdem der Taut vorliegt.... Wird der Erfolg so, wie ich bestimmt hoffe, würde ich mit Freuden Ihre verlegerische Idee aufgreifen und aus Ihren Arbeiten über die neue französische Architektur einen ähnlichen zweiten Band machen.*"

151. Archiv S. Giedion, letter from Giedion, 27 April 1928: "*Da es sich dabei um eine europäische Autorität handelt, so ist es von Vorteil, gerade gegenüber der Reaktion, wenn wir dies rückwärts vermerken können.*"

152. Archiv S. Giedion, letter from Giedion, 15 August 1927: "*...eine geschickte propagandistische Arbeit [ist] wichtiger als ein überstürztes Herausbringen.*"

153. Archiv S. Giedion, letter from Georg Biermann, 19 November 1927: "*...dürfte die neue Architektur in Frankreich so oder so im Titel nicht fehlen, mag bei Ihnen auch noch so sehr der eigentliche Nachdruck auf den Eisenbeton liegen.*"

154. Ibid.: "*Was die künstlerische Ausstattung anlangt, so sollten wir diesmal versuchen, ohne Künstler auszukommen, nachdem sich das Publikum tatsächlich an den beiden letzten Taut-Büchern zum Überdruss satt gesehen hat. Die Einfachheit und die Klarheit scheint mir der Stil zu sein, den die Zeit auch innerhalb der Buch-'Architektur' benötigt, und ich glaube, bei der buchtechnischen Qualität würde man auch ohne die Hilfe eines Künstlers, die unter Umständen sehr gefährlich ist, jede Garantie nach der angedeuteten Richtung hin übernehmen können.*"

155. Cf. Alfred Gellhorn, "Eisen, Eisenbeton," *Deutsche Allgemeine Zeitung*, 8 February 1928: "*Der Vortragende war, im ganzen genommen, ein Redner, wie man solchem vielleicht nur einmal im Leben begegnet. Seine ungeheuer knappe, mit drei Worten Ganzheiten gebende und das Wesen durchdringende Sprechweise allein bedeutet etwas Ausserordentliches.*"

156. Manuscript in the Archiv S. Giedion.

157. Dr. Ra, "Bauen in Frankreich," *Hannoverscher Anzeiger*, 15 February 1928.

158. "Tradition des neuen Bauens," *Neue Zürcher Zeitung*, 6 June 1928.

159. Archiv S. Giedion, letter from Giedion, 27 April 1928: "*Der Stoff des Buches ist sehr heikel und ich muss bei einem so umstrittenen Thema von der penibelsten Gewissenhaftigkeit im Ausdruck sein.*"

160. Archiv S. Giedion, letter from Giedion, 25 May 1928: "*Das gibt dem ganzen Kapitel einen andern Sinn. Ich möchte Sie bitten, keine Sinnänderungen vorzunehmen, ohne mich zu fragen, sonst kommen unwillkommene Dinge vor, die ich nicht verantworten kann.*"

161. A loose errata sheet inserted into the book notes, "*Auf Seite 6 muss die Überschrift*

***Architektur?** heissen*" (The subhead on page 6 must read **Architektur?**). I owe this reference to Werner Oechslin.

162. Giedion had suggested a simple sheet printed on both sides. See his letter to Biermann of 22 May 1928 in the Archiv S. Giedion.

163. Giedion's letter to the publishers, Klinkhardt & Biermann, Leipzig, of 25 May 1928 in the Archiv S. Giedion.

164. Archiv S. Giedion, letter from Giedion, 1 June 1928: "*Bestätige unsere beiden Telephongespräche. Nach Ihrem heutigen Gespräch wird nun der Umschlag und Einband lauten:*

BAUEN IN FRANKREICH

BAUEN IN EISEN

BAUEN IN EISENBETON

Der innere Buchtitel[:]

BAUEN IN FRANKREICH

EISEN

EISENBETON

. . . *Das Schlagwort in der Literatur muss lauten: BAUEN IN FRANKREICH.*"

165. C. O. Jatho, in *Literaturblatt, Beilage* to *Frankfurter Zeitung*, 21 October 1928: "*Wenn man, wie Giedion, die Akzente richtig setzt, kommt man hier zu nicht geringer Wertschätzung des vielgeschmähten 19. Jahrhunderts.*" Archiv S. Giedion, 43-T-3, Besprechungen-Rezensionen.

166. Karl With, in *Die Literatur, Beilage* to *Kölnische Zeitung*, 18 August 1929. Archiv S. Giedion, 43-T-3, Besprechungen-Rezensionen.

167. *Dresdner Zeitung*, 30 June 1928. Archiv S. Giedion, 43-T-3, Besprechungen-Rezensionen.

168. Alfred Gellhorn, in *Das Unterhaltungsblatt*, literary *Beilage* to *Deutsche Allgemeine Zeitung*, 20 September 1928. Archiv S. Giedion, 43-T-3, Besprechungen-Rezensionen.

169. Gustav Stotz, in *Werkbund Gedanken*, supplement to *Stuttgarter Neuen Tagblatt*, 25 August 1928: "*Giedion deckt . . . die Grundlagen einer Tradition auf, die für unser heutiges Bauen, dessen entscheidende Gestaltungen in Eisen und Beton geschaffen werden, als einzig echte bezeichnet werden darf. Durch die Art, wie er die Verbindungen und Zusammenhänge zwischen den konstruktiven Leistungen des vorigen Jahrhunderts mit den modernen auf dem behandelten Gebiet blosslegt, gewinnt das Buch eine ungeheure Aktualität und Lebendigkeit.*" Archiv S. Giedion, 43-T-3, Besprechungen-Rezensionen.

170. Curt Glaser, in *Berliner Börsen-Courier*, 7 July 1928. Archiv S. Giedion, 43-T-3, Besprechungen-Rezensionen.

171. Meyer (see note 136), 91: "*Nach der Lektüre dieses Buches wird niemand mehr die moderne Architektur als traditionslosen und traditionsfeindlichen Bolschewismus verächtlich machen können, noch behaupten, die nationale Nuance werde vom konstruktiven Bauen erstickt.*"

172. Gellhorn (see note 168): "*Weiter ausgesponnen wünschte man gern die nur angedeute-*

ten Anschlüsse an die grossen Konstrukteure Englands und Deutschlands, dann aber auch Amerikas."

173. Paul Joseph Cremers, in *Rheinisch Westfälische Zeitung* (Essen), 4 November 1928: "*Sehr viele Anzeichen deuten darauf hin, dass Deutschland die Führung übernehmen wird.*" Archiv S. Giedion, 43-T-3, Besprechungen-Rezensionen.

174. Paul Klopfer, "Funktion, nicht Form! Randbemerkungen zu Sigfried Giedions Buch *Bauen in Frankreich, Bauen in Eisen, Bauen in Eisenbeton*," *Stein Holz Eisen* 43, week 6 (1929): 86–89. Archiv S. Giedion, 43-T-3, Besprechungen-Rezensionen.

175. Gellhorn (see note 168): "*. . . erhält der erste Teil des Buches, der dem Bauen in Eisen gewidmet ist, die stärksten Aufschlüsse und Anregungen.*"

176. *Hoch + Tiefbau* (Zurich), 29 September 1928, 328. Archiv S. Giedion, 43-T-3, Besprechungen-Rezensionen.

177. Leo Adler, in *Wasmuths Monatshefte für Baukunst* 12, no. 9 (1928): 429–30. Archiv S. Giedion, 43-T-3, Besprechungen-Rezensionen.

178. Herman Sörgel, in *Die Kunst* (May 1929): 27–28: "*Wenn Giedion sagt: Dieses Durchspültsein des Hauses von Luft: innen, aussen, unten (!), oben, dies ist es, was wir vom neuen Haus verlangen, so kann man mit dem gleichen Rechte sagen: dies ist gerade das, wovor uns das Haus schützen soll, sonst brauchte man ja überhaupt kein Dach über dem Kopfe.*" Archiv S. Giedion, 43-T-3, Besprechungen-Rezensionen.

179. Max Osborn, in *Vossische Zeitung*, 11 November 1928: "*Knapper, einleuchtender und sinnfälliger (vorzüglich die beschrifteten Abbildungen) ist diese Entwicklung [das Werden der modernen Bauformen] noch nicht gezeigt worden.*" Archiv S. Giedion, 43-T-3, Besprechungen-Rezensionen.

180. Justus Bier, in *Nürnberger Zeitung*, 5 May 1930. Archiv S. Giedion, 43-T-3, Besprechungen-Rezensionen.

181. Adolf Behne, in *Die Welt am Abend*, 4 March 1929: "*Das Buch, reich an wichtigem neuem Material, typographisch durch Moholy famos besorgt, gehört zu denen, die unentbehrlich sind.*" Archiv S. Giedion, 43-T-3, Besprechungen-Rezensionen.

182. Archiv S. Giedion, letter from Paul Klopfer, 25 June 1928: "*Es war mir geradezu eine Genugtuung für meine Deduktionen von Ihnen die Versicherung zu erhalten, dass hinter die Architektur im üblichen Sinne ein Fragezeichen gehöre. Ich werde das Buch auch für meine Schule bestellen, denn es steckt ein grosser pädagogischer Wert darin. Auch meine noch etwas veralteten Lehrer dürften profitieren, und ich mit, denn sie glauben mir immer nicht.*"

183. Archiv S. Giedion, letter from Walter Benjamin, 15 February 1929: "*Sehr geehrter Herr Gidion*

als ich Ihr Buch bekam, elektrisierten mich die wenigen Stellen die ich las derart, dass ich mir vornehmen musste, nicht eher an die Lektüre zu gehen als bis ich den Kontakt mit meinen ihm verschränkten Untersuchungen in höherem Masse besässe als es im Zeitpunkt seines Eintreffens, äusserer Umstände wegen, der Fall war. Seit einigen Tagen sind nun bei mir die Dinge wieder in Fluss gekommen und ich verbringe Stunden über Ihrem Buch, in Bewunderung. Noch kenne ich es erst in seinem letzten Teile. Absichtlich schreibe ich Ihnen, solange ich die Bewegung, in die es mich versetzt, noch regiere. Ihr Buch stellt einen der wenigen Fälle dar, die wohl jeder kennt: dass wir vor der Berührung mit etwas (oder jemandem: Schrift, Haus, Mensch etc.) vorher wissen, dass sie im höchsten Grade bedeutsam geraten muss, dieses Wissen täuscht nicht.

Ich studiere an Ihrem Buch (neben so vielen anderen, in dem es mich aufs unmittelbarste angeht) die herzerfrischende Differenz von radikaler Gesinnung und radikalem Wissen. Sie haben das letztere, und darum sind Sie imstande, die Tradition aus der Gegenwart heraus zu erleuchten, oder vielmehr zu entdecken. Eben daher die Noblesse Ihres Werkes, die ich neben seinem Radikalismus am meisten bewundere. Mit Ihnen hierüber sprechen zu können, würde mich sehr erfreuen. Ich wäre Ihnen dankbar, wenn Sie bei einem gelegentlichen Berliner Aufenthalt sich meiner erinnerten. Sollte ich im Frühjahr nach Paris kommen, so werde ich mir erlauben, Sie zu benachrichtigen. Endlich nehmen Sie bitte mit den letzten Worten den Dank für die Zusendung *Ihrer Schrift, den ich bis hierher verschob.*

Mit den aufrichtigen Empfehlungen
15. Februar 1929
Berlin-Grunewald
Delbrückstraße 23

Ihr sehr ergebener
Walter Benjamin"
(emphasis in original).

Walter Benjamin consulted *Bauen in Frankreich* for his unfinished work, the so-called Arcade project (*Passagen-Werk*). There are excerpts from Giedion's book in his notes, particularly in connection with the subject of "iron construction." Of particular importance are the references Benjamin makes to *Bauen in Frankreich* in his epistemological notes: "*Genau so, wie Giedion uns lehrt, aus den Bauten um 1850 die Grundzüge des heutigen Bauens abzulesen, wollen wir aus dem Leben [und] aus den scheinbar sekundären, verlorenen Formen jener Zeit heutiges [Leb]en, heutige Formen ablesen*" (Just as Giedion taught us to read the basic features of modern architecture in the buildings from the middle of the nineteenth century, so we want to read modern life and modern forms from the life and from the apparently secondary, lost forms of that period). And: "*So hat auch der Historiker heute nur ein schmales, aber tragfähiges Gerüst—ein philosophisches—zu errichten, um die aktuellsten Aspekte der Vergangenheit in sein Netz zu ziehen. Wie er die grossartigen Ansichten, die die neuen Eisenkonstruktionen von den Städten gewähren—s[iehe] a[uch] Giedion Abb[ildungen] 61/63—auf lange hinaus sich ausschliesslich den Arbeitern und Ingenieuren erschlossen, so muss auch der Philosoph, der hier die ersten Aspekte gewinnen will, ein selbstständiger, schwindelfreier, wenn es sein muss einsamer Arbeiter sein"* (So, too, today's historian has to build up only a narrow but solid philosophical framework in order to pull the most topical aspects of the past into his net. Just as the splendid views that the new iron constructions lend to the cities – s[ee] a[lso] Giedion fig[ures] 61/63 – have for a long time been mastered exclusively by workers and engineers, so must the philosopher, who wants to get at the earliest points of view in this matter, be an independent, sober, if necessary, lonely worker). Walter Benjamin, *Das Passagen-Werk,* ed. Rolf Tiedemann (Frankfurt am Main: Suhrkamp, 1983), 1: 215f. and 572.

184. In the autobiographical postscript to his book *Architektur und Gemeinschaft: Tagebuch einer Entwicklung* (Hamburg: Rowohlt, 1956), 137, Giedion wrote: "*Lange Zeit zögerte ich, dieses nüchterne Büchlein HEINRICH WÖLFFLIN zu schicken. Schliesslich tat ich es doch. Zu meinem Erstaunen kam die Antwort zurück: 'Glauben Sie nicht, dass geheime Linien von* [*Renaissance und Barock*] *zu* [*Bauen in Frankreich*] *führen?'*" (For a long time I hesitated to send this sober little book to HEINRICH WÖLFFLIN. However, I finally did it. To my astonishment I received the answer: "Don't you think that secret lines lead from [*Renaissance und Barock*] to [*Bauen in Frankreich*]?"). Perhaps Giedion knew already in 1928 that with his book he had largely abandoned the ground of Wölfflin's tradition. In a letter of 27 April 1928 to his other Munich teacher, Paul Frankl, he wrote about his almost-completed book and said about Wölfflin: "*Wölfflin habe ich in der ganzen Zeit meines Hierseins ein einziges Mal durch Zufall gesehen. Er ist glaube ich ziemlich isoliert und ihm sind im Grunde alle Erscheinungen unter seinen Schülern verdächtig, die nicht unbedingt in Epigonenschuhen stecken. Darum habe ich ihn auch nie besucht*" (I have seen Wölfflin only once, by chance, during the entire time I have been here [i.e., back in Switzerland after his studies in Munich]. I think he is quite isolated, and he is basically suspicious of any efforts by his students that do not wholeheartedly follow in his footsteps. That's why I have never visited him). (Archiv S. Giedion, letter by Giedion, 27 April 1928). Nevertheless, in later years Giedion repeatedly referred to his art-historical training and to what was perhaps the most important part of it: Heinrich Wölfflin's instruction.

185. In a later letter Biermann corrected this number to eight hundred. Archiv S. Giedion, letter from Georg Biermann, 22 December 1928.

186. Archiv S. Giedion, letter from Georg Biermann, 12 December 1928: "*Der Absatz Ihres Buches ist trotz der Weihnachtspropaganda sehr minimal. Man ist in Leipzig ein wenig verzweifelt und kann garnicht erklären, warum ein Buch, das bereits so glänzende Kritiken hatte, so miserabel geht. Ich glaube nicht, dass bis heute 600 Exemplare fest verkauft sind. Meiner Überzeugung nach ist der Titel Schuld daran, und ich bedauere sehr, dass Sie mir da im letzten Moment noch in den Arm gefallen sind. Die Folgen habe ich vorausgesehen und kann es Ihnen schwarz auf weiss beweisen, dass man in einem solchen Fall den klugen Erfahrungen eines alten Verlegers nicht vorgreifen darf.*"

187. For the French edition Giedion was in contact with Le Corbusier, who acted as an intermediary for Editions Crès.

188. Archiv S. Giedion, letter from Georg Biermann, 5 January 1929: "*Glauben Sie mir, dass an dem Misserfolg Ihres Buches nur der törichte Titel Schuld hat. Kein Deutscher, nicht einmal ein deutscher Architekt, interessiert sich für die neue Baukunst in Frankreich, wohl aber für das wichtigste Problem, das Bauen in Eisen und Eisenbeton. Solange nicht dieser Titel fällt, nützen meinem Gefühl nach alle Anstrengungen nichts.*"

189. Less than a year later Frank Otten presented Giedion with a large part of the translation.

BUILDING IN FRANCE
BUILDING IN IRON
BUILDING IN FERRO-CONCRETE

Jacket of *Building in France* designed by László Moholy-Nagy based on the negative of a photo taken by Sigfried Giedion of the *pont transbordeur* in Marseilles. Courtesy Archiv S. Giedion, Institut für Geschichte und Theorie der Architektur, ETH-Hönggerberg, Zurich.

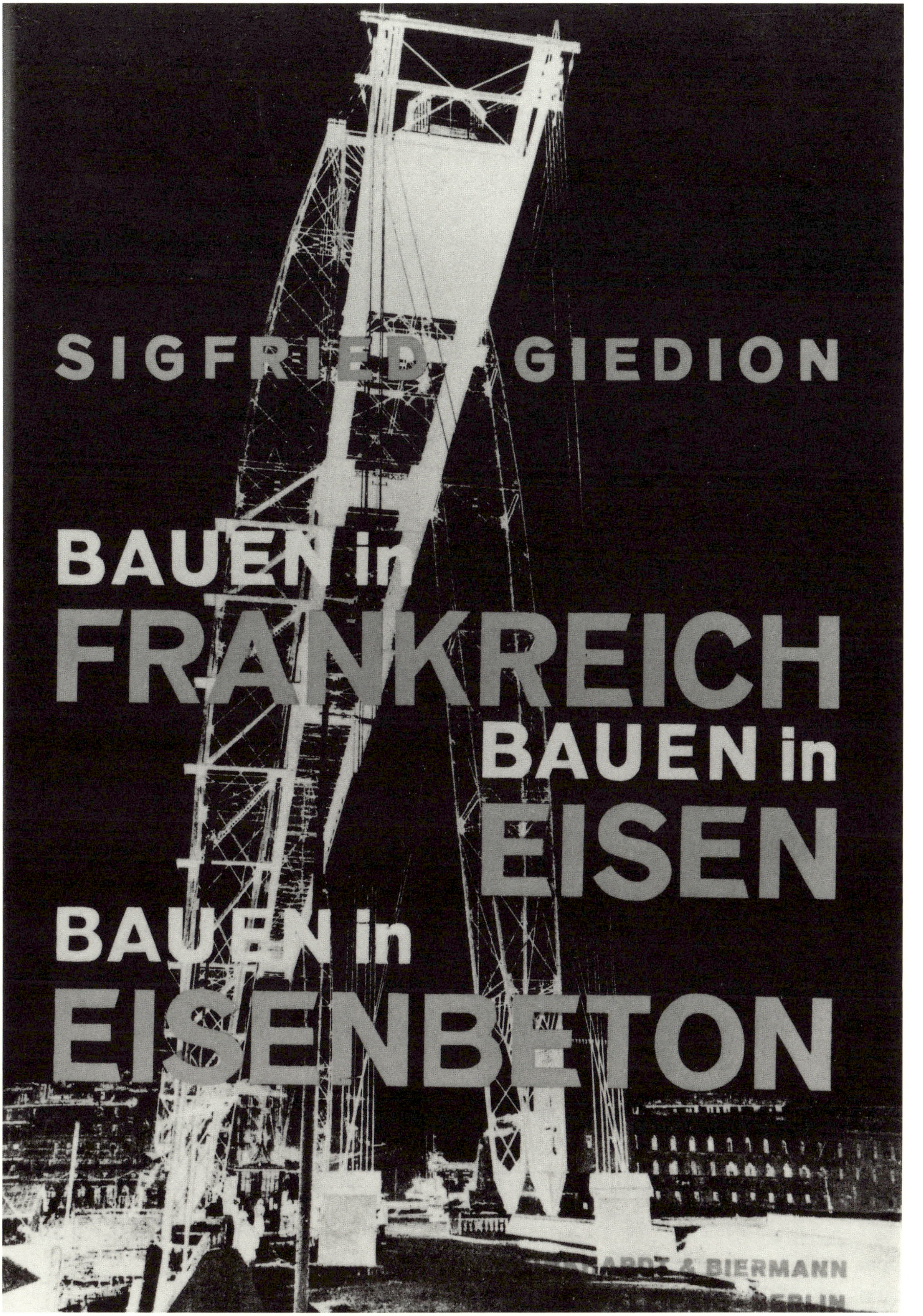
SIGFRIED GIEDION
BAUEN in
FRANKREICH
BAUEN in
EISEN
BAUEN in
EISENBETON
BIERMANN

PRELIMINARY REMARK

This book is written and designed so that

it is possible for the hurried reader to understand the developmental path from the captioned illustrations;

the text furnishes closer explication;

the footnotes provide more extensive references.

Jacket and cover: L[ászló] MOHOLY-NAGY, who also oversaw typography and layout.

This work was facilitated through the cooperation of the librarian of the Conservatoire des Arts et Métiers (Paris), through A[uguste and] G[ustave] PERRET, TONY GARNIER, LE CORBUSIER, and the other architects mentioned in the book.
I would particularly like to thank the excellent structural engineer of the Technische Hochschule Zurich, Prof. H[ans] JENNY-DÜRST, who carefully examined the page proofs. Unfortunately, due to the advanced stage of production, and partially because of the perspective from which the book was written, not every suggestion could be considered.

INTRODUCTION

Even the historian stands within, not above, time. He has lost the pedestal of eternity. It is the same with poets, musicians, and architects in relation to the general public; with parents and teachers in relation to children; and in the relation between man and woman: we struggle on equal levels.

Vital, forward-looking periods seem to have hardly a place for pedestals. We do not fear the past. Past, present, and future are for us an indivisible process. But we do not live looking backward; we live looking ahead. The past strengthens us, for it provides us with the certainty that our will is not individually confined. But the future, whatever it may hold, is more important to us. The first concern is with those things that are *becoming*: children are more defenseless than the old.

The *task* of the historian is first to recognize the seeds and to indicate—across all layers of debris—the continuity of development. The historian, unfortunately, has used the perspective of his occupation to give eternal legitimation to the past and thereby to kill the future, or at least to obstruct its development.

Today the historian's task appears to be the opposite: to extract from the vast complexity of the past those elements that will be the point of departure for the future.

In every field the nineteenth century cloaked each new invention with *historicizing masks*. In the realm of architecture as well as in the realm of industry or society. New constructional possibilities were created, but at the same time they were feared; each was senselessly buried beneath stone stage sets. The vast collective apparatus of industry was created, but every attempt was made to distort its significance in that only a few were allowed to benefit from the advantages of the production process.

This historicizing mask is inseparably linked to the image of the nineteenth century. It cannot be denied. But by the same token we must not forget the *forward momentum* that permeates the nineteenth century.

If we extract from that century those elements that live within us and are alive, we see with surprise that we have forgotten our own particular development—if you will, our TRADITION.
Brushing away the decades of accumulated dust atop the journals, we notice that the questions that concern us today have persisted in unsettled discussion for more than a century.
We see at the same time, indeed with greater assurance, that the architecture we now describe as "new" is a legitimate part of an entire century of development.
Since it belongs to a great stream of development, we must even refuse to see its origins in a small number of architectural precursors around 1900—for instance, [Hendrik Petrus] Berlage, [Henry] van de Velde, [Frank] Lloyd Wright, [Peter] Behrens, [Auguste and Gustave] Perret, [Tony] Garnier. The "new" architecture had its origins at the moment of industrial formation around 1830, at the moment of the transformation from hand work to industrial production. We scarcely have the right to compare our century with the nineteenth as far as the boldness of its advance and its works are concerned.
The task of this generation is: to translate into a HOUSING FORM [*Wohnform*] what the nineteenth century could say only in abstract and, for us, internally homogenous constructions. Everyone knows that we are therefore still at the very beginning of a long-neglected transformation from handicraft to industrial building production.
The division by generations that was necessary and that we ourselves have yet to complete may perhaps become more acceptable as we confront the past without prejudice. For we have no fear that it may yet crush or confuse us.

Frauenkirch-Lengmatt
February 1928

GIEDION

CONSTRUCTION

Is CONSTRUCTION something EXTERNAL?

We are being driven into an indivisible life process. We see life more and more as a moving yet indivisible whole. The boundaries of individual fields blur. Where does science end, where does art begin, what is applied technology, what belongs to pure knowledge? Fields permeate and fertilize each other as they overlap. It is hardly of interest to us today where the conceptual boundary between art and science is drawn. We value these fields not hierarchically but as equally justified emanations of the highest impulse: LIFE! To grasp life as a totality, to allow no divisions, is among the most important concerns of the age.
Physiologists have shown that a person's body build and nature are inseparably connected. Science traces specific characters back to certain bodily types. The connection between respiration and mental balance has been discovered. The body takes its form internally through breathing, gymnastics, sport. To overdevelop an arm muscle, or to douse the face with cosmetics like an isolated body (as the arteries harden), is no longer acceptable.
Construction is also not mere ratio.[1] The attitude that drove the previous century to expand our knowledge of matter, so much that it resulted in a previously inconceivable command of it, is as much the expression of an instinctive drive as is any artistic symbol.
We say that art anticipates, but when we are convinced of the indivisibility of the life process, we must add: industry, technology, and construction also anticipate.
Let us go further: architecture, which has certainly abused the name of art in many ways, has for a century led us in a circle from one failure to another.
Aside from a certain *haut-goût* charm the artistic drapery of the past century has become musty. What remains unfaded of the architecture is those rare instances when construction breaks through. Construction based entirely on provisional purposes, service, and change is the only part of building that shows an unerringly consistent development. Construction in the nineteenth century plays the role of the subconscious. Outwardly, construction still boasts the old pathos; underneath, concealed behind facades, the basis of our present existence is taking shape.

1) We mean here not just the creative intuition that every great constructor must have. It is well known that he fixes the dimensions mostly emotionally and that calculation often comes later only as a test. We mean construction itself, which is not determined by purpose alone, but which seems also to transcend rational values and is expressive. This also challenges the old prejudice that art and construction may be neatly divided, by presenting art as "unintentional" and "purposeless," and construction alone as "purposeful."

INDUSTRY

Industry completes the transition from handicraft to machine production.
Industry is only part of the problem connected with the transition from individual to collective design.
Machine work means serial design, precision. Handicraft has its own special charm that can never be replaced: the uniqueness of the product.
But without machine work there is no higher technology. By hand one can neither mill sprocket wheels that fit frictionlessly together, nor draw out uniform wire, nor profile iron precisely. The transition from individual to collective design is taking place in all fields, practical as well as spiritual ones.
Now, it is the case that INDUSTRY, which is intensively involved with the life process, displayed this change before other fields—private life or art—took note of it.
Industry, big industry, is a result of the French Revolution.[2]
The Assemblée Nationale initiated its development with the *Proclamation of the Liberty of Labor* of 2 March 1791.
With this proclamation of free competition the guild system (*les corporations*) was at once abolished.[3]
Before the French Revolution articles for everyday use were produced by the guilds. Guild membership was just as limited as the number of workers or helpers each member could take on and the kinds of product each could produce. That meant privilege in favor of a few and an extraordinary burden (*gêne onéreuse*) on the consumer. The complex instrument of industry was created through the possibility of a free division of labor.
Like construction, industry is an inner expression of the life process.
Though we are objectively able to create anticipatory designs, old mental "residues" prevent us for a long time from drawing the human consequences:
INDUSTRY anticipates society's inner upheaval just as construction anticipates the future expression of building.
Even before industry existed in its present sense—around 1820—Henri de Saint-Simon (1760–1825)[4] understood that it was the central concept of the century and that it was destined to turn life inside out:
"The whole of society rests upon industry."
It seems that the force of Saint-Simon's influence on the schools and tendencies of the century lay, above all, in his ability to grasp the emerging reality and to transform it into a utopia. It is the opposite method to the cultural idealism that dominated Germany at the time, which neglected reality in order to pursue emanations of pure spirit.

2) There were several industrial and joint-stock companies already under the ancien régime. [Charles] Ballot, *L'Introduction du machinisme dans l'industrie française* (Paris, 1923), p. 23, discusses the epoch from 1780–92 and another 1792–1815, which, under the effect of the Revolution, introduced the machine to a few areas (cotton, wool). Industry in today's sense was first introduced around 1830.
3) Davioud, "Un discours d'architecte," *Encyclopédie d'architecture,* 1878, p. 27.
4) *Système industriel,* 1821. *Catéchisme des industriels,* 1823.

Saint-Simon foresaw the great concentrations of labor, the urban centers, and the factories with thousands of workers that transferred the results of research directly into action. As a consequence of an industrial economy he foresaw the dawn of a classless society, the end of war, and the end of national borders: a single army of workers spanning the globe. The end of man's exploitation of man (*l'exploitation de l'homme par l'homme*) will have been achieved. The eye of the visionary no doubt simplifies and leaps over intermediary stages:[5] Saint-Simon never reckoned with the century's divided soul, which in architecture as in society imposed the old formal apparatus on the new system.

The anonymous process of production and the interconnected procedures that industry offers only now fully take hold of and reshape our nature.

[5]) Saint-Simon, himself rooted in the feudal system, only formulated the elementary contradictions of the military and industrial system. His students quickly drew the consequences of his system.

Fig. 1. PONT TRANSBORDEUR (1905) and HARBOR of MARSEILLES
A mobile ferry suspended by cables from the footbridge high above the water connects traffic on the two sides of the harbor. This structure is not to be taken as a "machine." It cannot be excluded from the urban image, whose fantastic crowning it denotes. But its interplay with the city is neither "spatial" nor "plastic." It engenders floating relations and interpenetrations. The boundaries of architecture are blurred.

ARCHITECTURE[1]

The concept of architecture is linked to the material of stone. Heaviness and monumentality belong to the nature of this material, just as the clear division between supporting and supported parts does.
The great dimensions that stone requires are for us still habitually connected with each building. It is entirely understandable that, with their unusually modest dimensions, the first buildings executed in tensile materials time and again evoked among contemporaries the concern that the building might collapse.
Architecture is linked to the concept of "monumentality." When the new building materials—iron and ferroconcrete—assume the forms of gravity and "monumentality," they are essentially misused.
It seems doubtful whether the limited concept of "architecture" will indeed endure.
We can hardly answer the question: What belongs to architecture? Where does it begin, where does it end?
Fields overlap: walls no longer rigidly define streets. The street has been transformed into a stream of movement. Rail lines and trains, together with the railroad station, form a single whole. Suspended elevators in glazed shafts belong to it just as much as
Fig. 1 the insulating filling between the supports. The antenna has coalesced with the struc-
Fig. 59 ture, just as the limbs of a towering steel frame enter into a relationship with city and

Fig. 2. EIFFEL TOWER (1889)
Interior of pier
Instead of a massive tower, an open framework condensed into minimal dimensions. The landscape enters through continuously changing snippets.

harbor. Tall buildings are bisected by rail lines. The fluctuating element becomes a part of building.

Architecture has been drawn into the current from the isolated position it had shared with painting and sculpture.

We are beginning to transform the surface of the earth. We thrust beneath, above, and over the surface. Architecture is only a part of this process, even if a special one. Hence there is no "style," no proper building style. Collective design. A fluid transition of things.

By their design, all buildings today are as open as possible. They blur their arbitrary boundaries. Seek connection and interpenetration.

In the air-flooded stairs of the Eiffel Tower, better yet, in the steel limbs of a *pont transbordeur,* we confront the basic aesthetic experience of today's building: through the delicate iron net suspended in midair stream things, ships, sea, houses, masts, landscape, and harbor. They lose their delimited form: as one descends, they circle into each other and intermingle simultaneously.

One would not wish to carry over into housing this absolute experience that no previous age has known.[6] Yet it remains embryonic in each design of the new architec-

[6]) A fascination of Corbusier's houses consists in the fact that he has attempted this as much as possible.

Fig. 3. PETROLEUM TANK, CONCRETE BRIDGE, STREET, TRESTLE (MARSEILLES)

Fig. 4.
The same industrial landscape. Chimneys of a sugar refinery in the background. The various traffic levels, the juxtaposition of objects determined only by necessity offer — so to speak unconsciously and as raw material—possibilities for how our cities may later be designed openly without the constraints of preestablished levels.

ture: there is only a great, indivisible space in which relations and interpenetrations, rather than boundaries, reign.
The concept of architecture has become too narrow. One can no longer contain, like radium in a bottle, the need to create that which is called art and explain what remains of life devoid of it.
The ponderous movement of human affairs has as its consequence that the new attitude toward life manifests itself much sooner in the objective fields—such as construction, industry—than in those fields that lie close to us.
Only now is the housing form being seized by those hidden forces that a century ago drove man to the constructional and industrial attitude.
Our inner attitude today demands of the house:
Greatest possible overcoming of gravity. Light proportions. Openness, free flow of air: things that were first indicated in an abstract way by the constructional designs of the past century.
Thus, the point is reached where building falls in line with the general life process.

CONSTRUCTOR AND ARCHITECT

The immense influence of the Ecole Polytechnique (established 1794/95) during the first three decades of the nineteenth century can be attributed to the fact that for the first time the task was consciously posed to establish a connection between science and life, to bring about a connection between higher mathematics, physics, and applied technology.
Subsequently this extended to penetrating life constructionally instead of with the craft experience. In all fields.
Jean-Antoine Chaptal, the great chemist, industrialist (and minister under Napoleon) already clarified this goal at the beginning of the nineteenth century, for he thought science should climb down from its pedestal in order to lend a hand in the work of easing the difficult path through reality.
Since [Jean] Rondelet, the theorist and preserver of the Panthéon in Paris, insisted in his "Address on Starting a Course on Construction at the Ecole Spéciale d'Architecture" (1816) that construction must be considered in a completely different way than previously, the engineer has increasingly encroached upon the field of the architect.
At the same time the encroachment of the engineer-constructor signifies the encroachment of more rapid, industrial means of design, which first offered the possibility: to create the foundations of contemporary life.
Unconsciously, the constructor assumes the role of a guardian in the nineteenth century: by continuously pressing new means upon the architect, he keeps the latter from altogether losing himself in the vacuum.
The constructor presses for a design that is both anonymous and collective. He renounces the architect's artistic bombast. Upsets his special position. That is his function.
Since the advance of industry around midcentury, we sense how the "artist-architect" feels his privileged position menaced, out of which grows the increasingly spasmodic emphasis on his "artistry."
This concern reached its apex with the development of industry. After 1890, when industrial development had lost the wonder from its time of origin and had become self-evident, the threatening influence of the constructor abated. Underground it continued.

Instead of derivations, some voices from various moments of the period:

1850:

> We will create a characteristic architecture of our own at the moment we make use of the new means offered by the new industries. The application of cast iron permits and enforces many new forms, as we can observe in railroad stations, suspension bridges, and in the vaults of conservatories.

The Romantic Théophile GAUTIER 1850. Journal *La Presse*.

1864:

Can one claim that the public is pleased when one hears complaints daily and sees how often it prefers engineers to architects? Why this partiality? It is simply because engineers do not take inflexible positions but satisfy themselves with the strict fulfillment of the given program (*remplir rigoureusement le programme*), whereas architects, all too often advancing what they call "beauty," violate the legitimate requirements and needs of the patron.

Anatole de BAUDOT,[7] *Réorganisation de l'Ecole des Beaux-Arts, de son influence sur l'étude de l'architecture* (Paris: chez A. Morel & Co., 1864), p. 5.

1867:

"Is it the fate of architecture to give way to the art of engineering (*génie civil*)? Will the engineer one day absorb the architect?" One speaks of the organic art of the future, and at the same time it is clear what the present predicament is: "Where does eclecticism lead? An eclectic atmosphere completely envelops the modern world; all organs of respiration absorb it and, mixed with our blood, it acts on heart and brain."

(César DALY, *Revue générale de l'architecture*, 1867, p. 6.)

1877:

The institute arranged a competition on the theme: "The union or the separation of engineers and architects." [Gabriel-Jean-Antoine] DAVIOUD, the architect of the Trocadéro, received the prize with the response: the union of architect and engineer must be indissoluble. "The solution will be real, complete, fruitful only when architect and engineer, artist and scientist, are fused together in the same person. We have for a long time lived under the foolish persuasion that art is a kind of activity distinct from all other forms of human intelligence, having its sole source and origin in the personality of the artist himself and his capricious fancy...."

(*Encyclopédie d'architecture*, 1878, p. 67.)

1889:

For a long time the influence of the architect has declined, and the engineer, *l'homme moderne par excellence*, is beginning to replace him. Were the engineer able to replace the architect altogether, the latter could undoubtedly disappear without at the same time eradicating art.

Forms will not compose the basis of the new architecture. In the general disposition of plans and in the design of constructional systems arising from these new situations is to be found the new expression as a whole; the details will then follow.

7) In 1863–64 an effort was made to reform the estranged teaching methods of the Académie des Beaux-Arts. A. de Baudot's first publication appeared on this occasion.

But you will say what you propose is indeed the method of the engineer. I do not deny it, for it is correct.

Anatole de BAUDOT,[8] to the International Congress of Architects, 1889.

1 8 8 9 :

It is not in the studios of the painters and sculptors that the revolution so long awaited is preparing—it is in the factories![9]

The novelist Octave MIRBEAU in *Le Figaro,* 1889.
(Cf. *Encycl. d'arch.,* 1889/90, p. 92.)

1 9 2 4 :

The century of the machine has awakened the CONSTRUCTOR; new tasks, new possibilities, and new means gave birth to him. He is at work now everywhere.

LE CORBUSIER, in the journal L'ESPRIT NOUVEAU, no. 25.

[8]) Anatole de Baudot, 1834–1915, is one of the most consistent pioneers of the new architecture. For decades he fights resolutely for an organic basis for architecture, his talent being more that of a prophet and propagandist than a practicing architect.
Like Labrouste, whose student he was, Baudot wishes to impart, first of all, an architectural education that was not predisposed to impracticality.
He sees in iron and later in ferroconcrete the appropriate building materials for the age. Indeed, one might speak of him as the first to have employed ferroconcrete practically in architecture: in
Fig. 77 his church of Saint Jean de Montmartre [Saint-Jean-l'Evangéliste], Paris, begun 1894.
As the leading force of the excellent journal *Encyclopédie d'architecture* he is often found formulating the guidelines for building with an unerring precision that we have yet to match. In his formulations, he is less restrained by aestheticism than Corbusier, for he comes directly from the school of Auguste Comte's rationalism and lived before the congestion [*Verschleimung*] caused by the arts and crafts.
For twenty-five years he defends his opposition to the academy and its corrosive influence—beginning in 1887—in his annual courses at the Trocadéro ("Cours d'histoire de l'architecture française"). Every year the course changed. The first lectures appeared in the *Encyclopédie d'architecture,* 1888/89. After his death Henri Chaine published them under the title: A. de Baudot, *L'Architecture, le passé—le présent* (Paris, 1916). Much of the original punch has been lost. One can learn from Baudot how to profit from the past without being crushed by it.
Baudot was a civil servant, a member of the Commission des Monuments Historiques, eventually rising to be its vice president. Baudot, a student of Labrouste and Viollet-le-Duc, thus had as his main occupation the preservation of monuments, but his creative force lies in the fact that he grasped the past functionally, not formally.

[9]) *"Pendant que l'art cherche l'intimisme ou s'attarde aux vieilles formules, le regard encore tourné vers le passé, l'industrie marche de l'avant, explore l'inconnu conquiert des formes"* [While art seeks intimacy or clings to outworn formulas with its eyes fixed on the past, industry forges ahead, explores the unknown, conquers new forms].
"Ce n'est point dans les ateliers des peintres et sculpteurs que se prépare la révolution tant prédite et tant désirée: c'est dans les usines!" [Translated in text, above.]

CONSTRUCTORS

The names of the constructors who gave shape to the nineteenth century are for the most part unknown. Just as in the Middle Ages, the actual development occurred anonymously. A few names that are related to our remarks are presented below.

It is noteworthy that the first wave—the pioneers—was born around 1800, and the second wave—those who completed the task—was born around 1830.

Many fit the picture of industrialists sketched by Saint-Simon. They were by no means narrow specialists. They embraced the new possibilities in all fields. Often they were also entrepreneurs.

The first wave—comprising those born around 1800—shares a prophetic and fanatic conviction in the realization of a worldview that existed nowhere but in the mind, in the vision.

Antoine POLONCEAU (1778–1847). A precursor. 1797 Ecole Polytechnique. Built roads through Alpine passes (Simplon 1801–06, Mont-Cenis 1812–14) and experimented with the use of concrete in foundations. Erected one of Paris's most beautiful cast-iron bridges (pont du Carrousel, 1839). At the same time he was concerned with the rationalization of agriculture, improvements in drainage, harvesting, and the soil.

Eugène FLACHAT (1802–1873). Railroad builder. Constructor. With Emile Péreire[10] he overcame the enormous resistance of both the government and the populace and in 1837 built the first railroad in France (Saint-Germain). One should not forget that in 1834 a French minister, upon his return from England, declared in the Chamber of Peers: railroads would be an unrealizable dream, useful at most for the immediate environs of Paris, for Sunday excursions. Flachat made designs for docks, canals (Paris-Le Havre), warehouses, harbor facilities, transatlantic steam shipping, and an Alpine tunnel (1860). From this universal circle of ideas arose, along with blast furnaces, the designs for Les Halles of Paris. Flachat participated in all of the great industrial issues, but at the same time he was concerned with such details as the wear of train rails and the organization of large technical societies.

The industrial movement appears to have been so much in the blood of this generation that, for instance, Flachat built his house on a parcel of land between two tracks and the ceaseless whistles of trains.

Hector HOREAU (1801–72). Designer who did not achieve the goal but who formulated the decisive building tasks before it was possible to realize them on such a scale. He wants buildings that correspond to the needs of the age. He designs enormous exhibition halls (1837), libraries, market halls.

Like Flachat he demonstrates how one could design Les Halles of Paris (1845) in the new materials, but [Victor] Baltard builds them. In 1850 Horeau receives the first prize for the design of the London Industrial Exhibition; one month later John

10) The banker Emile Péreire (born 1801) works closely with the Saint-Simonists. Auguste Comte (born 1798), who witnessed the dawning of the "positive," or scientific, age, belongs to this generation, as do other important pupils of Saint-Simon, [Amand] Bazard (born 1791) and the diplomat [Ferdinand de] Lesseps (born 1805), who initiated the Suez and Panama canals.

Paxton wins the commission.[2] Misfortune follows him to the end. The existence of such prophets is certainly not without significance. It encourages the subsequent realization of the vision.

Henri LABROUSTE (1801–75). Attempts for the first time to combine engineer and architect in one person: architect-constructor.

Grand Prix de Rome at twenty-three. Unstifled after five years at the Villa Medici. Sees antiquity organically instead of sentimentally and views his Italian sojourn (considered the highest reward) as a systematic alienation from life, "for after his return the *pensionnaire de l'état* is faced with an entirely different reality."[11] Labrouste designs no romantic-classicist royal palaces.

Labrouste wins prizes for prisons (Alexandria), hospitals (Lausanne), and, after waiting twelve years for the commission: the library of Sainte-Geneviève! The methods he uses are the methods of direct observation, those "positive" methods upon which Auguste Comte simultaneously "wanted to establish the lawfulness of social life."

Labrouste designs a plan "according to climate, material, and the necessities of the program."[12] He recognizes the function of iron before railroads course through France; he sees construction as the innermost face of architecture—the exterior only as an encasement (*enveloppe*), or epidermis—at a time when the methods to control it were still remote.

Among those in the second wave who are able to lead construction to completed designs—besides Camille POLONCEAU (1813–47) or HENRI DE DION[13] (1823–78)—is Gustave Eiffel, the only one whose name is not forgotten. Gustave EIFFEL (1832–1923), engineer-constructor, entrepreneur. In 1858 he works on the large iron bridge near Bordeaux and attempts to sink pylons with the then-new method of compressed air. His life is occupied above all with one great problem: the arched girder, the arched truss (*ferme en arc*). He attempts to solve this problem with all the graphic, mathematical, and experimental expertise available. The arched truss, which springs lightly in a parabolic curve from one bridge support to the next, signifies release from the rigid relationship of unilaterally supporting and supported members: for the first time in 1869 with the viaduct over the Sioule, in 1875 with the bold bridge over the Douro, and in 1879 with the viaduct over the Garabit. (Span of the parabolic support 165 meters.) His *tour de 300 m* (1889) displays in its foundations four such bridge supports translated into architectonic form. In addition, there are also warehouses, railroad stations, and the locks of the Panama Canal. The fields overlap.

11) Henri Labrouste, "Travaux d'élèves de l'Ecole de l'architecture de Paris," *Rev. gén. de l'arch.*, 1840, p. 543.

12) Ibid.

13) The great Berlin constructor J[ohann] W[ilhelm] SCHWEDLER (1823–91) also belongs to this second wave. He is the genuine theoretician and calculator of the so-called spatial framework (shallow dome of the Berlin Gasworks, 1863). He built the great Vistula bridges and the arrival hall of the Frankfurt am Main railroad station.

THE NINETEENTH CENTURY

Like every age, the nineteenth century is a complex whole. Perhaps more confused than other ages. We are still too close for conclusive judgments.
We should first of all establish two currents that—often inseparable—interpenetrate: first inheritance: to it belong all fields that somehow stand apart. Art, celebration, representation, private affairs.
Second: life itself compels its own laws. The surface of the earth is transformed as never before. (Harbors, railroads, tunnels, great thoroughfares.) Production established on a totally different basis (industry). The regrouping of society begins (socialization tendencies).
Inheritance is a part of us. It should not be neglected. Yet it ends bitterly if it assumes priority over emerging life, if it violates it. This is the case in the nineteenth century. Indeed, in the long run, such usurpation is only mock rule; subliminally, life compels form, but there emerges an oppressive atmosphere that time and again demands upheaval.
This is especially obvious in the field of architecture: if we employ the usual stylistic notions to understand it, we are left with nothing but empty shells. We might say that throughout the last century one built with a bad conscience or with uncertainty. Especially the best feel that one lives in a contaminating air of eclecticism. The nineteenth century's general attitude, torn by its urge toward integrated constructional design and its urge toward individual isolation, is necessarily pessimistic.

Again and again one tries to find a "style" without realizing that these formalistic experiments were condemned to failure from the start. Surface frills. The age of delimited styles based on handicraft ended decisively at the moment when the notion of an isolated architecture became untenable.
The nineteenth century: strange interpenetration of individualistic and collectivistic tendencies.
Like hardly any age before, all actions were labeled "individualistic" (the ego, Nation, Art), but underground, within disdained everyday fields, it had to create the elements of collective design, as in a frenzy.
Today everything rests on these elements.
We must concern ourselves with this raw material: with gray buildings, market halls, warehouses, exhibitions. However unimportant they may appear to be for the aesthetic titillation:
In them lies the kernel!

THE NINETEENTH CENTURY IN FRANCE

The role of France is well established in the painting and literature of the nineteenth century. This is not at all as clear with architecture. The academic incrustations bear the blame. They dazzled all formally educated souls. When the new architecture will have advanced far enough to allow a broader survey, it may become evident: all the academic incrustations were unable to smother the constructional soul of French architecture!
France played the leading role in the history of nineteenth-century constructivism.

From the beginning of the century, two poles opposed one another in France: L'Ecole des Beaux-Arts—L'Ecole Polytechnique.
Napoleon founded the Ecole des Beaux-Arts in an unfortunate moment and thereby revived an institution of the ancien régime. By that, official architecture became a branch of the fine arts. That became its ruin. In the Baroque, this unity had been complete and self-evident. But in the course of the nineteenth century it had become conflicting and false. Even today the academy des Beaux-Arts proves itself to be a most distressing drag on active development.
Around the Ecole Polytechnique gathered the great mathematicians ([Gaspard] Monge, [Joseph-Louis] Lagrange),[3] physicists, constructors, economists, and Saint-Simonists. Time and again, up to 1830, we see the blue uniform of the *polytechniciens* in the pictures of the Revolution. The crucial contact between theoretical and applied science emanated from this school.

Paging through the architectural journals of the century, one sees that the two questions that most preoccupied contemporaries grew out of the dissension between these two schools:
1. Along what lines should the training of an architect proceed?
2. What is the relationship between engineer and architect? How are their rights distributed? Are they one and the same?
All other questions are of a secondary nature, formal disputes.
Hence the struggle of the academy against the Gothicists or the struggle of the Gothicists against the new "bastard style": the Neorenaissance.[14] Similarly, the current striving for so-called national styles basically has a different meaning. National differences develop through the influences of climate, material, and formative will, utterly independently and unconsciously. The struggle toward a "national" style, with its desire to retain formal-handicraft details, is fought like the struggle of the Gothicists or Renaissancists, on formal rather than on functional grounds. Screened from the real events.

[14]) Cf. Paul Léon, "Les querelles des classiques et gotiques," *Rev. de Paris* (July 1913).

IRON

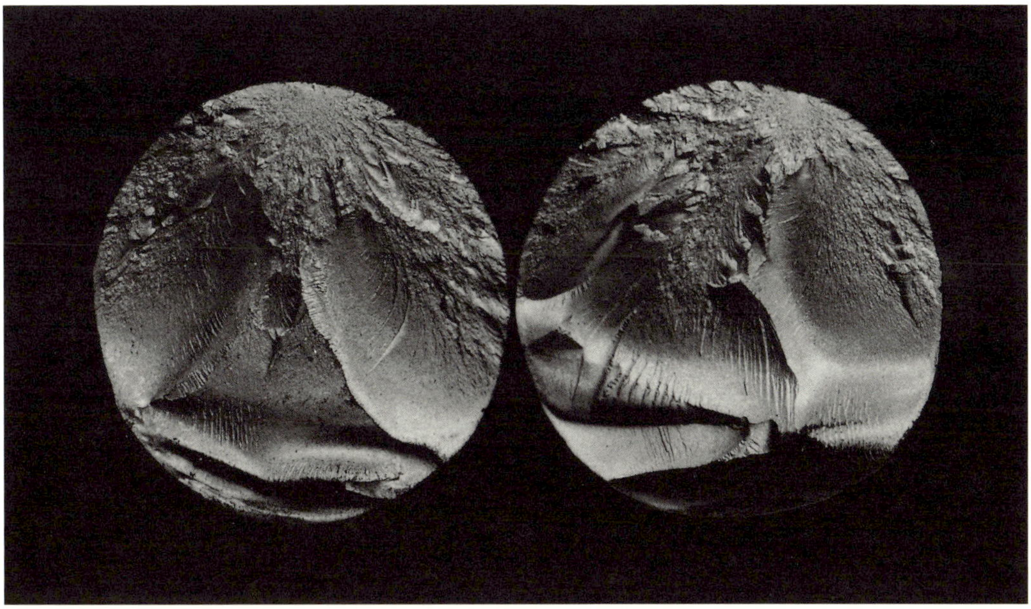

Fig. 5. STEEL. Broken Steel Rod Enlarged Three Times
Note the great density of the structure in contrast to other building materials, such as wood: from this molecular structure arise certain properties and types of construction.

The introduction of iron into architecture signifies the change from craftsmanship to industrial building production. The beginnings of the new architecture can be dated to the day when the old production methods were abandoned and mechanically manufactured rolled iron replaced handwrought iron.

STONE can only resist compression. It allows inert masses to be layered into piles, but only the most extreme inventiveness allows it to achieve a certain hollowing out (Gothic). Stone signifies compact mass. Stone massively closes off spaces. The entire width of a wall is load-bearing. Broad horizontal openings contradict its structure.

IRON can be stretched and drawn together. It resists extension and pressure and hence bending.
The significance of iron is: to condense high potential stress into the most minimal dimensions. If a comparison is permitted, iron suggests both muscular tissue and skeleton in a building. Iron opens the spaces. The wall can become a transparent glass skin. To design a load-bearing wall becomes an intolerable farce.
This leads to new laws of design.

Instead of the rigid balance of support and load, iron demands a more complex, more fluid balance of forces. Through the condensation of the material to a few points, there appears an unknown transparency, a suspended relation to other objects, a creation of the airspace, *des combinaisons aériennes* that Octave Mirbeau recognized already in 1889. This sensation of being enveloped by a floating airspace while walking through tall structures (Eiffel Tower) advanced the concept of flight before it had been realized and stimulated the formation of the new architecture. Not through superficial formal derivation but rather by an inner law.

Of all the new building materials, iron has the longest tradition of construction. It is fascinating to see with how much fanaticism the fight for its introduction persisted for half a century, and with what regularity its design slowly revealed itself. We can follow this gradual realization and discern its slow growth.[15]

Demand for its introduction existed before the machine made possible the manufacture of iron profiles. Even before the means were found to calculate dimensions theoretically, vision and ratio recognized it as the material of the future.

This went so far that at an early point in time—1849—the HOUSING FORM on which we are today working was already anticipated:

"Glass is destined to play an important role in METAL ARCHITECTURE. Instead of thick walls, whose strength and solidity are diminished by a great number of openings, our houses will be so permeated with openings that they will appear translucent. These wide openings of thick, single- or double-glazed glass panes, either frosted or transparent, will allow a magical splendor to stream in during the daytime, stream out at night."[16]

It seems that in the human organism certain attitudes develop that only later crystallize in reality. Just as industry was recognized as the fulcrum of the century before the development of mechanical engineering, so, too—on a reduced scale—is a housing form today emerging before the social structure is ready for it.

[15]) By coincidence I received a copy of A[lfred] G[otthold] Meyer's EISENBAUTEN (Esslingen, 1907) only when this book went to press. One will find much supplementary material there. The different point of departure is naturally explained by the different year of publication. Nevertheless, we would like to note that our approaches often fortunately concur.

[16]) Jobard, "Architecture métallurgique," *Rev. gén. de l'arch.*, 1849, p. 30: "*... Nos maisons seront émaillées d'élégantes et nombreuses ouvertures qui les rendront perméables à la lumière.*"

FIRST ATTEMPTS

It started with the introduction of iron in roof framing. The wooden beams of theaters and warehouses burned like tinder. One tried to replace them with iron. Soon one saw that iron construction required little space, allowed much light to stream in, and, when used in combination with glass, was especially suited for the roofing of courtyards. Glass and iron galleries appeared, the true point of departure for railroad stations, market halls, exhibition buildings. Before the introduction of rolled-iron beams, complicated experiments with composite iron-and-wood systems were conducted to make roofs fireproof, to render them *incombustibles*.[17] **Fig. 6** **Fig. 9**

In addition to purely technical application (bridges, railroads) there were building details: such as cast-iron stairs and iron canopies in theaters (apparently introduced by [Jacques-Ignace] Hittorff and Lecomte[4] in the Théâtre de l'Ambigu-Comique after 1824).

Glass houses,[18] with their — compared to walls — virtually invisible exterior shell, provide the impetus for the introduction of cast-iron supports and skeleton constructions.

England, industrially far ahead of France, enjoys the lead until about midcentury.[19]

Visible exponent: the Crystal Palace in London, the Industrial Exhibition of 1851.

From the beginning, France showed itself to be superior in construction.

The books by Ch[arles-] L[ouis-Gustave] Eck document more fully how varied the efforts were regarding the introduction of iron already in the century's first three decades.[20] [5] The architect L[ouis-] A[uguste] Boileau's books give a good insight into the next three decades.[21] [6]

The first building that anticipates subsequent developments is the library of Sainte-Geneviève (1843–50) by Henri LABROUSTE. Henri Labrouste is without doubt the most prominent figure in the field of architecture at the beginning of industrial development. As a *pensionnaire de l'Académie* in Rome he did not approach antiquity in an aesthetically ceremonious way but scrutinized the temples of Paestum for "the organism of each construction."[22]

17) Ch. L. Eck, *Traité de l'application du fer, de la fonte et de la tôle dans les constructions civiles, industrielles et militaires, dans celles des ponts fixes ou suspendus, des écluses et des digues de la mer* (Paris, 1841).

18) One of the first glass houses built completely with an iron structure is found in the Jardin des Plantes, Paris 1833. Cf. M. Neumann, *L'Art de construire et de gouverner les serres* (2nd ed., Paris, 1846).

19) English constructions are bold and ponderous, such as the enormous cast-iron roof truss that Mathieu Clark designed in 1833 for the Théâtre Alexandrin in [Saint] Petersburg. (Illus. in Ch. Eck, *Traité de construction etc.*, Paris 1841.) Our **fig. 15** from the same period shows one simply replaced wood with cast iron.

20) Ch. Eck, *Traité de construction en poteries et fer* (Paris, 1836). Ch. L. Eck, *Traité de l'application du fer, de la fonte et de la tôle* (Paris, 1841).

21) L. A. Boileau, *Le fer, principal élément constructif de la nouvelle architecture* (Paris, 1871).

22) Henri Delaborde, "Notice sur la vie de Henri Labrouste. Lue dans la séance publique annuelle du 19 octobre 1878," p. 8.

It seems typical that nothing substantial has been written about a master like Labrouste since his death and the related eulogies, despite the fact that he was one of the most significant architectural per-

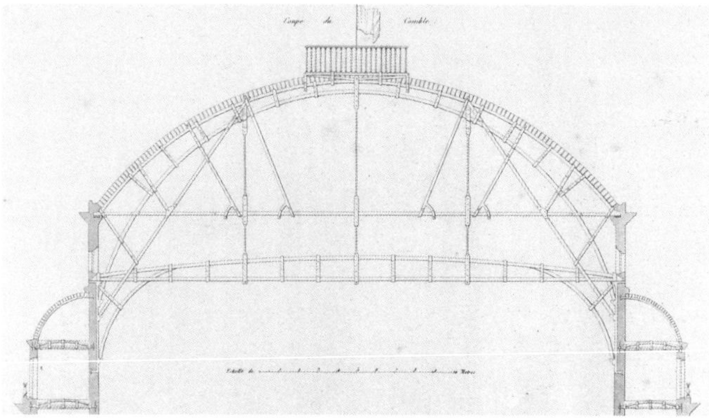

Fig. 6. Victor LOUIS: Wrought-Iron Truss of the Théâtre-Français. 1786
It seems remarkable to us how—in 1786—the reinforcement of the truss instinctively follows the correct moment of inertia, which was studied only later:[23] "What is most noteworthy in this roof is the thinness of the walls on which it rests. . . . Finally, the ingenious form of buttressing applied to the weak points in the structure by external vaults in iron and clay tile" (Ch. Eck, *Traité,* p. 50).

Fig. 7. [François-Joseph] BÉLANGER and [F.] BRUNET: Corn Exchange (Paris). 1811
The architect Bélanger and the engineer Brunet gave the Halle au Blé its complicated iron-and-copper structure in 1811 (the wooden dome of 1783 burned down in 1802). To our knowledge this marks the first time that architect and engineer were no longer combined in one person. (Cf. Brunet, *Dimensions des fers qui doivent former la coupole de la Halle aux grains* [Paris, 1809]). Hittorff, the builder of the Gare du Nord, received his first insight into iron construction from Bélanger. At any rate, here it behaves more as an application of iron than as an iron construction. The methods of wooden construction are simply transferred to iron.

sonalities of the nineteenth century. In the course of this work, when we wanted a better look at the original plans for the Bibliothèque Nationale, it became evident that even these have disappeared. It would have been important finally to bring documentary clarity to the academy's sabotage-like conduct toward Labrouste and his students. In principle, it is still true today that the academy sabotages and suppresses young talent by preventing the realization of their work only to assimilate them in old age.

23) A few years earlier, Victor LOUIS at least partially attempted something similar in his large theater in Bordeaux. Ill. in [Sigfried] GIEDION, *Spätbarocker und romantischer Klassizismus* (Munich, 1922), p. 184.

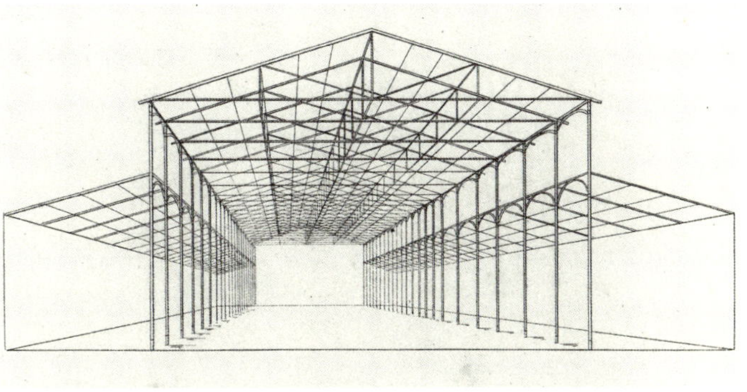

Fig. 8. [M.-G.] VEUGNY: Market Hall of the Madeleine. 1824
The gracefulness of the slender cast-iron columns recalls Pompeian wall paintings. "The wrought- and cast-iron construction of the new Madeleine market hall is one of the most graceful productions of this kind; nothing more elegant or in better taste could be imagined..." (Eck, *Traité*, ibid.).

Fig. 9. [Pierre-François-Léonard] FONTAINE: Galerie d'Orléans in the Palais-Royal (Paris) 1829–31
Even FONTAINE, one of the founders of the Empire style, converted in old age to the new material. In 1835–36 he replaced the wooden floor of the Galerie des Batailles in Versailles with iron panels. Galleries like those in the Palais-Royal underwent further development in Italy. They represent for us the point of departure for new building problems: railroad stations, etc.

Fig. 10. Henri LABROUSTE: LA RÉSERVE
(Ground floor of the library of Sainte-Geneviève.) Cast-iron columns lead freely and boldly through the space as a part of the iron skeleton sunk into the building.

He demonstrated that although students of the academy produced beautiful drawings of antique details, they completely missed the inner organism of the building. He learned to recognize "that the best buildings from an artistic standpoint were precisely those constructed by the simplest, most truthful, and most rational methods."[24] For the first time, he expressed an expanded MEANING of CONSTRUCTION, yielded by the new possibilities:[25] The essence of construction is found not in the isolated study of the mason's or locksmith's handcrafted details but in the interpenetration of every part of a building.
Labrouste belongs to the generation of 1830 that, as noted in a completely different connection, was guided by a single grand current, by the demand for renewal of social, moral, and intellectual life.[26]
By the time he was assigned the library, Labrouste was regarded by everyone as the purest incarnation of the *esprit nouveau*.[27] For twelve years he had roamed Paris without being entrusted with even a single building. Labrouste was past forty when he received the library commission.

24) Eugène Millet, "Henri Labrouste," Extrait du *Bulletin de la Société centrale des architectes*, Exercice 1879–80, p. 5.
25) This is one of the rare personal statements of the architect H. Labrouste: "*la construction consiste dans la combinaison de toutes les parties architecturales*" [construction consists of the combination of all the architectural parts] "Travaux des élèves de l'école d'architecture de Paris pendant l'année 1839," *Revue gén. de l'arch.*, 1840, p. 59.
26) Willy Spühler, *Der Saint-Simonismus* (Zurich, 1926), p. 22.
27) Delaborde, p. 13.

1926

Fig. 11.
LE CORBUSIER: COOK HOUSE
It took approximately eighty years before one dared to show uninhibitedly the freedom of Labrouste: an uninterrupted construction (column) even in a living space.

Science and industry gave him very little assistance. Nevertheless, with the library of Sainte-Geneviève he attempted for the first time to insert an iron skeleton into a building, from the foundation to the roof. Sainte-Geneviève is at the same time the first pure library building in France. Labrouste was more sensitive to the possibilities of iron than his architectural contemporaries. The material corresponded with his intention: to condense the meaning of all things![28]

Labrouste inserts the iron frame into the building like the works into a clock:

The massive masonry core encasing the building still remains unaffected, but within this masonry core, from the ground floor to the ridge of the roof, is placed an iron system: columns, ceilings, vaults, girders, roof construction.

In individual rooms of the ground floor (*la réserve*), cast-iron columns without visible beams are connected to the upper story. These slender cast-iron tubes run down the middle of the room, attached to the ceiling only by a narrow flange. Sleek function, no beam with the hint of support and load, no ornament, no capital. These are things that today are dared only by a Corbusier or a Mart Stam. **Fig. 10** **Fig. 11**

The upper story, a double-aisle reading room (84 meters long, 21 meters wide), forms a single structural skeleton with the roof. The semicircular trusses are supported by cast-iron columns and—along the walls—consoles. If the plans are correct, Labrouste already split these semicircular ceiling trusses into three segments so as not to make them totally rigid, in order to allow for expansion. As we know, it was more **Figs. 13, 14 [12, 13]**

[28]) Delaborde, p. 13.

Fig. 12. H. LABROUSTE: LIBRARY OF SAINTE-GENEVIÈVE. 1843–50
Section through the upper story (reading room) and the roof truss.

Fig. 13. H. LABROUSTE: LIBRARY OF SAINTE-GENEVIÈVE
Reading room with cast-iron truss. The windows are placed high to allow continuous shelving.

than four decades later (Galerie des Machines, Exhibition 1889) that such a line of reasoning led to a real form: three-hinged arches. Figs. 47 to 52 [47, 48, 50–53]

The roof framing rests atop these semicircular trusses and the columns. Upon entering the attic space, one is astonished by the bold dimensions and by the slender, ingeniously stiffened wrought-iron struts carrying the heavy zinc roof. Labrouste achieved an astonishing thinness in the barrel vault by spanning the trusses with an iron webbing bound in plaster. One cannot help thinking of the delicate reinforcement of Perret's eggshell-thin concrete vaults in Casablanca and Le Raincy. Figs. 75, 76

Naturally, Labrouste could not use the Polonceau truss, the recently invented visible tie system, for a library. He prepares his own system with primitive means, going straight to the point that subsequently leads to completely new structural form: iron construction has to be balanced in itself!

Labrouste's contemporary, the theoretician Léonce Reynaud, succinctly summarizes this: "The iron skeleton, which results in the suppression of the vault's lateral thrust, is here realized in a perfect system."[29]

In his BIBLIOTHÈQUE NATIONALE (commissioned 1857, reading room and stacks opened 1867), Labrouste perfects this system. Again a tall aqueduct-shaped system of arches encloses the space. The iron framework of cast-iron columns stands free of the walls. The free suspension and self-supporting frame are thus clearly emphasized.[30]

[29]) Léonce Reynaud, *Traité d'architecture* (Paris, 1852–56), p. 387: "*L'ossature ferronnière, qui a pour conséquence la suppression de la poussée des voutes, se trouve formulé là dans un système complet.*"

[30]) Unfortunately, it is not possible to explore this more fully here. The thoroughgoing functional logic of the building acquires a very new meaning not only in the reading room with its domes that should provide each desk area with uniform light but also in the stacks where perhaps this effect is even more intense. Cf. *Revue gén. de l'architecture,* 1878, p. 144.

FRENCH CONSTRUCTION

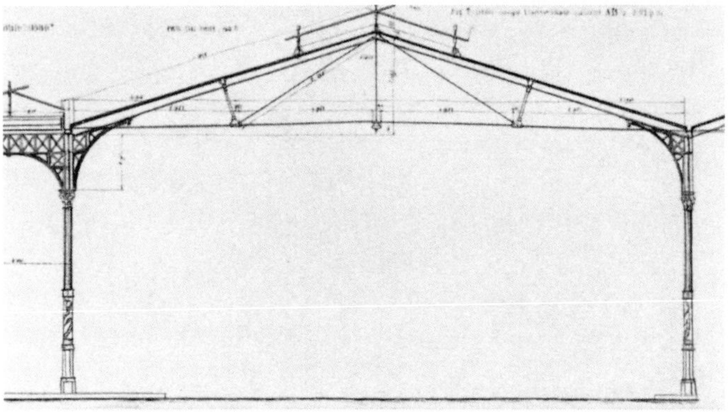

Fig. 14. POLONCEAU TRUSS from the Gare du Nord. Arrival Hall (1862)

The Polonceau truss, invented around 1837, on the occasion of the first French railroad, already displays careful attention to the inner structure of the material iron. RESULT: lightness of construction, elimination of any lateral thrust. Only the weight of the truss bears down on the supporting surface, therefore thin supports, very clear layout of the spaces. The visible tie-rods are still disturbing. One will try to eliminate them. As soon as this has been accomplished constructionally (cf. fig. 51), we stand before unprecedented designs.

FIRST FORMATION

Fig. 5 In order for iron to conform in its building parts and shapes to its actual molecular properties three conditions were necessary:

1. Wrought iron and steel should be manufactured in such a way as to avoid the accidental flaws endemic in the handicraft method of production. The Englishman Henry BESSEMER brought about the decisive revolution. He prepared steel in a purely chemical way (through mineral decomposition and regulation of the air supply in a Bessemer converter), precisely mandated by decarbonization. 1855.

2. The most important step toward industrialization: mechanical production of certain FORMS (profiles) from wrought iron or steel. The fields overlap: one did not start with building members but with train rails. In an English rolling mill in South Wales[31] the American [Robert Livingston] Stevens turned out the first broad-base rail mechanically produced (through rolling), that is, rails with a wide foot and a narrow head—the kind still in use today. 1832. Here is the starting point for sectional iron, that is, the basis for structural frameworks. Iron skeleton.[32]

3. Taking into account the molecular properties of iron, science had to study the material's specific laws, and constructors had to find a formative process that differed from the treatment of wood. The engineer Camille

[31]) Cf. [A.] Haarmann, *Das Eisenbahngeleise* (Leipzig, 1891), p. 53. To roll is to stretch the red-hot piece of iron between ungrooved cylindrical rollers.

[32]) The new production methods slowly made their way into industry. Double-tee iron joists appeared as ceiling beams in Paris in 1845 because of a masons' strike and as a result of the high cost of timber during a building boom and the increasingly large spans. L. A. Boileau, *Le Fer* (1871), p. 65.

ENGLISH CONSTRUCTION

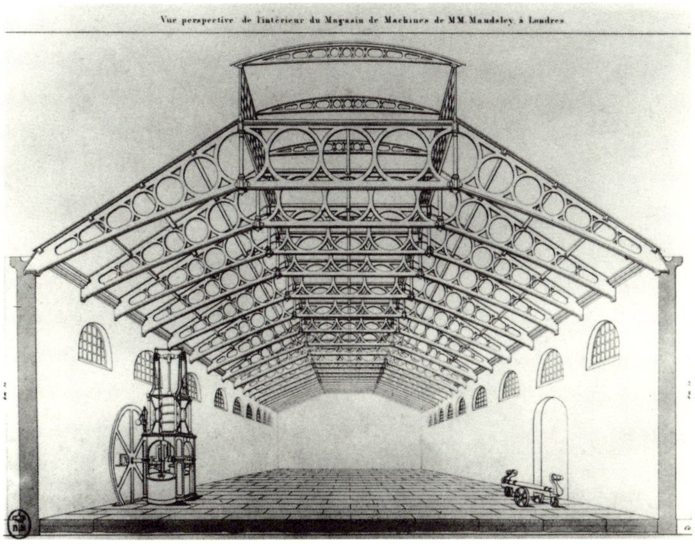

Fig. 15. [M. M. Maudsley] ENGLISH MACHINE HALL. 1833
Direct translation of the wooden roof truss into cast iron. RESULT: heavy construction and a need for thick walls to bear the lateral thrust.

POLONCEAU made the first steps in this direction. The Polonceau truss, which is still in use today, employs the material's preeminent property: its tensile strength. English cast-iron roof structures worked like Gothic vaults with enormous lateral thrust on the load-bearing walls. The significance of the Polonceau truss is that a light system of spanning members is self-contained, resting only with its own weight on the support, thus eliminating lateral thrust.[33] Fig. 14

The elementary guidelines are seen in buildings without prototypes in the past. Buildings shaped the new demands: metropolis, traffic, industry.
The common characteristic of these buildings is that they serve transient purposes: market halls, railroad stations, exhibitions.
Add to this—after 1870—the factory and the large warehouse as iron-skeleton construction.

MARKET HALLS

Metropolis problem of the nineteenth century entirely of a transient character. Function: in a few hours goods must daily be brought to and removed from a population of millions: *Les grandes Halles* of Paris.[34] Figs. 16, 17

33) In the first issue of the *Revue générale de l'architecture,* 1840, Camille Polonceau explains the reasoning behind his construction. He insists that every construction have: (1) durability; (2) economy; (3) the least possible dimensions; (4) the greatest simplicity. It is no accident that this construction was first used in an engine shed for the first railroad line Paris-Versailles. (Span 8.5 meters.)

34) In the excellent essay of César Daly ("Les halles," *Revue générale de l'architecture,* 1854) his judgment is thoroughly functional and farsighted. Of particular interest is the passage where Daly talks about the construction of Horeau and Flachat and withholds judgment because the material seems still too new for him to know for sure what form it should be given.

Fig. 16. BALTARD: LES HALLES (PARIS) at the Pavilion. 1853

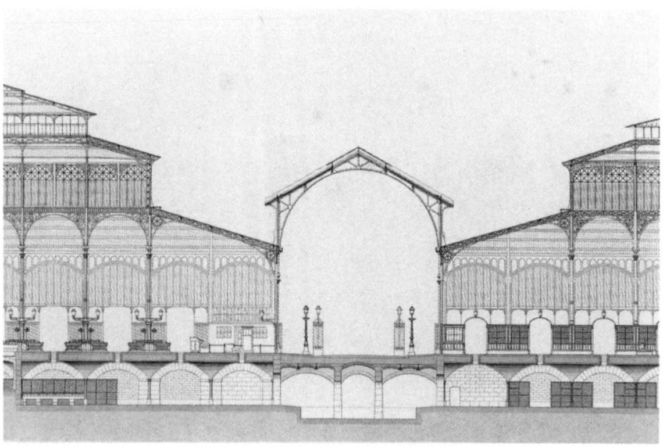

Fig. 17. BALTARD: LES HALLES (PARIS). 1853

Section through two pavilions and the connecting covered alley.
The markets consist of two groups of pavilions that are connected to each other by *rues couvertes*. This is a case of somewhat timid iron construction that avoids the ambitious spans of Horeau and Flachat and obviously clings to the model of the greenhouse.

What should a market hall accomplish?[35] "It must protect from the worst inclemencies of weather, besides it must consist of open halls with freely circulating air. This prevents stagnating odors and facilitates the delivery of goods."
What is required? "Freedom of movement, fresh air (*le grand air*), light, broadest prospect."[36] This demand for a CLEAR LAYOUT [*Übersichtlichkeit*] and the possibility of quick MOBILITY, which only the new materials could provide, are common to all the new building problems. To ascertain at what point we now stand in this development, consider the fact that the same requirements architects then used for the design of the individual new building tasks are today applied, with the same words, to the design of the entire city. Herein lies the leap from the then-acute management of passenger traffic to

[35]) *Revue gén. de l'arch.*, 1854, p. 22.
[36]) *Revue gén. de l'arch.*, 1854, p. 22.

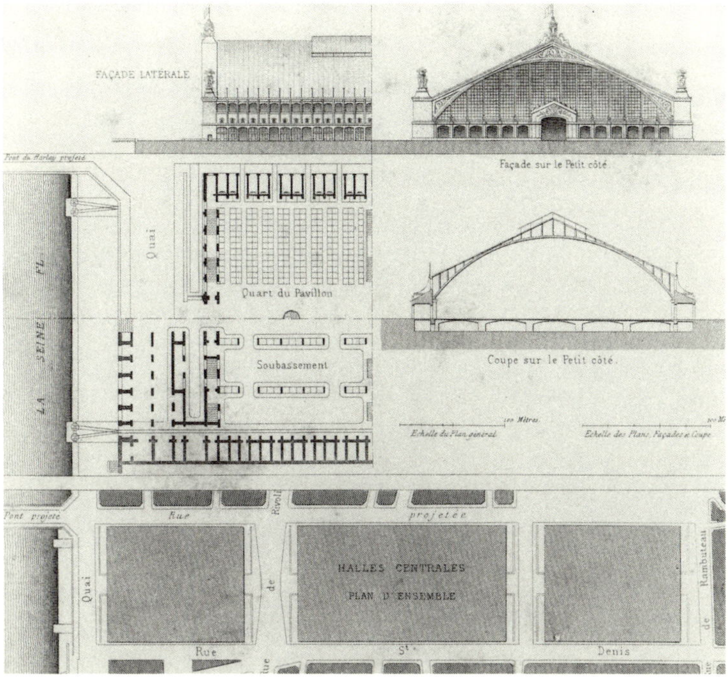

Fig. 18. HOREAU: DESIGN FOR LES HALLES. 1849

HOREAU proposed a span (86 meters) with enormous glass walls on the sides, a feat that was not realized until decades later. Extensive basements—the carriages were to be conveyed to these *soubassements* by elevators—were intended for the arriving goods. Horeau displayed his first designs in 1845. His large-scale design was shown in the Salon of 1849. We see in such projects what might be called anticipated development.

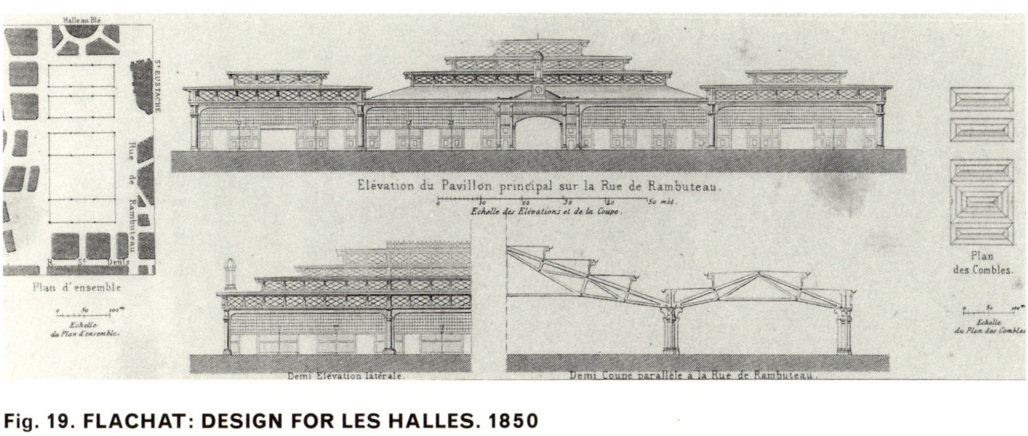

Fig. 19. FLACHAT: DESIGN FOR LES HALLES. 1850

This design with its Polonceau truss also had an audacious span (80 meters). Possibility of immediate execution with the available technical means. Unknown tautness of the freely spanning horizontal members and more open, clearer layout of spaces with a minimum of supports.

the now-acute management of the means of transportation (automobiles).
Victor Baltard (1805–74), whose name is still associated with Les Halles of Paris, was neither a great architect nor a greater constructor.
He first built a massive stone pavilion in 1851–52 and, after a trip to England, came to the conclusion that this country, too, had nothing new to offer. He had the support of the city administration of Paris, which, as L. A. Boileau says, envisioned a monumental style only in stone.[37]
But there were others in France who understood completely how to handle such a situation. First among them were Horeau, the architect, and Flachat, the railroad builder. Their plans and those of others had long been available when Baltard erected his stone pavilion.
Fig. 18 Horeau never executed a building, but he foresaw and anticipated the development. His plan for Les Halles shows a boldly sweeping parabola with a span of 86 meters. Although girded with masonry, it would have needed massive masonry piers to counter lateral thrust.
Fig. 19 Flachat's project was actually the basis for Baltard's later building. Yet its matter-of-fact solution is infinitely freer than Baltard's execution. It is altogether consistent with the development that a prophet of railroad construction would find the most elegant and practical solution: he was accustomed to designing necessities without inhibitions. These broad, spanning halls with Polonceau trusses were based on the idea: "Eliminate the solid to the utmost in favor of the void...increase the one at the expense of the other."[38] Condensing the supporting structure to a few points; avoiding columns as much as possible, hence vast spans.
Baltard's stone pavilion was already completed when the interested parties criticized
Figs. 16, 17 the *fort de la halle* in which "the vegetables and they themselves were enclosed in a citadel."[39] The prefect [Georges-Eugène] Haussmann ordered the pavilion pulled down. Baltard laboriously patched together his work with the ideas of others. [Félix] Narjoux[40] handed down an anecdote indicative of Baltard's position. As Napoleon III visited Baltard's new project under the direction of the prefect Haussmann, he asked, "Is it possible that two such contradictory projects stem from the same architect?" Haussmann countered that "the architect is indeed the same but the prefect is different." Despite Baltard's lack of originality,[41] the construction of Les Halles nevertheless contributed to the erection of hygienic and airy buildings of the same type even in the remotest provinces.

RAILROAD STATIONS

Their function, disposition, as well as transient nature compel the interrelation of architect and engineer.

[37]) L. A. Boileau, *Le fer, principal élément constructif* (Paris, 1871), p. 65: "The city councillors of Paris, subjected to morals and customs, could not imagine a monumental style otherwise than in masses of stone, and so they had selected a project designed accordingly." Precisely Baltard's first project.
[38]) *Revue gén. de l'arch.*, 1854, p. 31.
[39]) *Revue des deux mondes,* 1874 [no. 2], p. 803.
[40]) F. Narjoux, *Bâtiments élevés par la ville de Paris 1850–1880,* vol. 9 (Paris, 1883), pp. 11, 12.
[41]) Baltard also brought out a large folio publication on Les HALLES (Paris, 1863). Baltard's church of Saint-Augustin, Paris, 1860–71, employed a visible iron vault inside, yet this decoration had nothing to do with the design.

Fig. 20. HITTORFF: GARE DU NORD begun 1862
The glazed canopy is from a later date (1880).

Begun in 1862, the Gare du Nord (Paris) is the great work of this period. Better than its successors does it satisfy its function: the swift dispatch of traffic.[42]
The architect Hittorff's facade in the style of Roman public baths is soon forgotten. In the tightly measured ticket halls one can already see the combination of visible iron skeleton and stone. Immediately beyond, the expansive arrival halls open up. (Constructors: Couche and Boucher.) Engineers are responsible for the extraordinarily clear layout of the arrangement, the maximum of freely disposable space with a minimum of material, the airiness of the halls, whose grand arrangement still satisfies today's needs. The luxuriant abundance of space in the waiting rooms, entrances, restaurants, as seen around 1880, which led to the formulation of the railroad station problem as exaggerated Baroque palaces, is still completely avoided here.

DEPARTMENT STORES

The DEPARTMENT STORE is the emporium of industrial production. Like the covered market hall, the railroad station, the exhibition hall, it had no models available from the past. Like them, the department store is based on rapid service, on large-scale operation, on movement.
The idea for the *docks à bon marché* arose from the exhibition of 1867. The direct

[42]) The architect Hittorff (1792–1867) was concerned throughout his life with the new material iron and was probably well suited to work alongside constructors. More interesting than his well-known church of Saint-Vincent-de-Paul (begun 1832) were his "Rotonde du Panorama" (1838) and the "Cirque de l'Impératrice" on the Champs-Elysées that were probably the models for the overscaled rotunda of the Vienna World Exhibition of 1873. This rotunda, the work of an English constructor, is one of the first examples of how one can misuse iron by treating it as a surrogate for stone. The attempt to surpass the exterior dimension of Michelangelo's dome of Saint Peter's is quite evident.

Fig. 21. L. C. BOILEAU and EIFFEL: BON MARCHÉ DEPARTMENT STORE. 1876

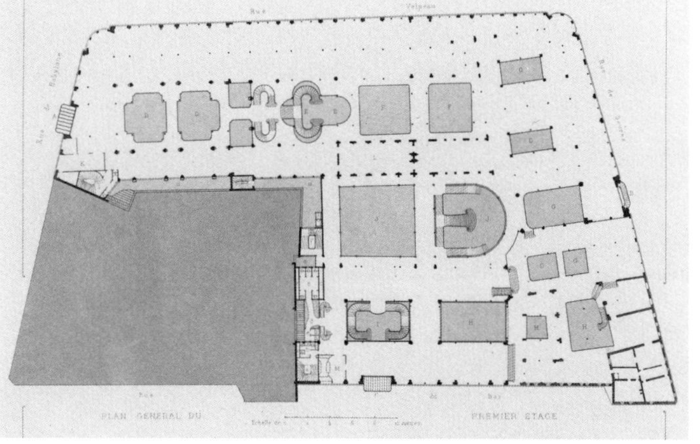

Fig. 22.
BON MARCHÉ DEPARTMENT STORE. Plan
First iron-and-glass department store. Surface area: 10,000 square meters.

Fig. 23. L. C. BOILEAU and EIFFEL: BON MARCHÉ DEPARTMENT STORE. 1876. Glass Roof
When the nineteenth century feels itself unobserved, it becomes bold. If possible, one conceals the new designs. Only gradually do the unobserved rear fronts of railroad stations, factories, the unspoiled forms of iron and concrete become visible.

impetus was the great disparity that everyone could see there between wholesale and retail prices. This price margin had to be reduced in order to meet the purchasing power of the poorer population.

The design of the department store demands:

Greatest possible freedom for circulation, clear layout,
Greatest possible influx of light.

Glass and iron thus become the constituent materials. *Glass* for the generous skylights as well as for the broad plate-glass windows for displays and the upper stories (side lighting).

The *iron skeleton* allows *thin* pillars within: freedom of circulation, clear layout, and it permits the best utilization of light at the front.

The *first* consistent realization of a department store in glass and iron is the MAGASINS AU BON MARCHÉ (Paris). 1876. *Eiffel* as engineer-constructor, L[ouis-] C[harles] *Boileau*, the son of the indefatigable advocate for the introduction of iron, as architect. **Figs. 21 to 24**

Broad windows at the front. Only the corner pavilions, a reminiscence of French châteaux, could not be abandoned, as was even the case later (Paul Sédille's Printemps). But already Boileau remarks that these stone pillars are no more than the

Fig. 24. Boileau and EIFFEL: BON MARCHÉ DEPARTMENT STORE, 1876, under construction[43]
Surprising lightness and unerring precision in this first attempt to solve the department-store problem. Like the Gare du Nord, still free of the later degeneration.

hors-d'oeuvre de la construction,[44] pure veneer (*placage*). The problem of the department store's interior is already understood, so that in principle the solution is retained in later cases: the different stories form a single space. One grasps them, "so to speak, all in one glance."
"Under no circumstances does such a building tolerate thick supporting walls, even massive piers are amiss here. Only pillars of small diameter are permissible; safety requirements have to be satisfied with these extremely limited elements." (Boileau).

43) The photos of the Bon Marché and the exhibitions of 1878 and 1889 are unpublished and are original to the period. Photo: Chevojon.
44) L. C. Boileau, *fils*, "Les magasins au Bon Marché," *Encyclop. d'arch.*, 1880, p. 184.

Fig. 25. Paul Sédille: Printemps Department Store (1881–89)
Here for the first time the fateful principle was taken seriously: "To attract the crowds and through decoration to seduce them into staying."[45] Sédille lets the iron frame rest on concrete foundations.

Fig. 26. Paul Sédille: Printemps Department Store (1881–89)
As with the Bon Marché, sheathing the skeleton with stone facings and tacking on pavilions.

Fig. 27. FRANTZ JOURDAIN: SAMARITAINE DEPARTMENT STORE (1905)
Increase of open space, overgrown with Jugendstil ornamentation inside and out, yet the framework is clean and does not lack the courage to display unclad iron also on the exterior.

[45]) Cf. *Science et l'industrie*, no. 143 (1925), p. 6: "*Accueillir la foule et la retenir en la séduisant.*"

EXPERIMENTAL ARCHITECTURE

EXHIBITIONS[46]

Almost every age, according to its own inner attitude, seems to develop a specific building problem: the Gothic the cathedral, the Baroque the palace, and the early nineteenth century with its nostalgic inclination to imbibe the past, the museum. In no building of German Romantic-Classicism is the will of the age more clearly manifested in form and idea than in [Karl Friedrich] Schinkel's Altes Museum in Berlin (1819). The Frenchman Henri de Saint-Simon simultaneously perceives industry as the central concept. Later in the century, between 1850 and 1890, as industry achieved its full development, exhibitions become the creative exponents of building production.[47] Only when industrial development's initial wonder had lapsed into a self-evident fact did the attraction and meaning of these arrangements abate.

Exhibitions are light buildings, quickly assembled and quickly disassembled: laboratories for industrial building.

One wanted to build in iron because stone, as a material, was far too heavy and expensive: "With iron, the various parts of the construction could be fabricated in the many separate workshops." Beyond this, iron had the advantage "of giving the building a special character, highly appropriate to its purpose." (See *Monographie. Palais et constructions diverses de l'exposition universelle de 1878, exécutées par l'administration*, p. 7.)

The history of exhibitions becomes the history of iron construction. Following the first tentative efforts of 1851 (London) and 1855 (Paris), it happened almost regularly that previously untried solutions on which a group of constructors was working were realized for the first time. Immediately thereafter, they left their stamp on life to the broadest extent. Often it was a matter of a daring and even dubious way of building (Eiffel Tower) into the unknown.[48]

Exhibitions not only summarized the results of the development but they also anticipated it.

In the history of exhibitions one can trace directly the transformation of the old static feeling of load and support into a new system of suspended equilibrium. Let us advance step by step to make this evolution clear.

[46]) The French word EXPOSITION—like the word INDUSTRIE—is much more ambiguous than the [German] word *Ausstellung* [exhibition]. *Exposition* also means: overview, juxtaposition, comparison, site, and even, in the figurative sense: representation of a theory.

[47]) A certain reservation is perhaps in order, for architecture in the nineteenth century—understood as a precursor of later evolution—was seen as an isolated field that could be grasped only with difficulty. In the crucial areas of design the fields overlap. Command of a collective formative will connects the exhibitions with the bridges, railroad buildings, machines, vehicles, in short, with the entire field of anonymous design.

[48]) This is confirmed by those works that are far removed from the exhibitions. In 1883 F[élix-Benjamin] Lucas says in the large publication *Les travaux publics de la France*, vol. *Routes et ponts*, p. 90: "Exhibition halls from 1855 and 1867 familiarized constructors with large spans and so increased the ideas of the specialized professionals that these dared, without hesitation, to erect a single-arch bridge with a span of 160 meters (bridge over the Douro)."

In an age of bounding transformation from handicraft to machine production exhibitions served to place all new discoveries next to one another immediately. Work proceeded feverishly and simultaneously in many places. Exhibitions now lined up the products of every country next to each other for comparison so that a mutual adjustment and intensification of the total production process happened as quickly as possible.

The exhibitions since 1850 were the first confirmations of a thoroughly global commerce. Yet their very existence indicated that no breaks are possible in the field of human activity.

Temporary and transitory in their very essence these arrangements were closely connected with life, and at the same time they were the birthplace of today's advertising. These exhibitions, mounted in a short time and at great expense, produced an as-yet-unknown intensification in the medium of publicity.

The whole range of human labor was to be embraced: all disciplines and, often even retrospectively, all periods. From agriculture, mining, from industry, from machines shown at work, to raw materials, to processed materials, to fine and applied arts.

This is due to a remarkable need for a premature synthesis that was also typical of other fields in the nineteenth century—the synthesis of the arts [*Gesamtkunstwerk*]. Apart from unquestionably utilitarian reasons, there was also the intention to give rise to a vision of the human cosmos in a new state of movement.

The exhibitions were born with industry. Like it, they were a result of the French Revolution. In the first half of the century they were limited events. Toward mid-century, as industrialization marched forth, the world exhibitions begin.

The first exhibition, which called itself *Première exposition des produits de l'industrie française,* took place in 1798 (year VI) on the Champ de Mars. It no longer dealt with luxury items—as did some presentations of the eighteenth century—but with articles for everyday use, such as clocks, safety locks, wallpapers, textiles, cotton yarns "carded and spun by machines" (page 51 of the exhibition catalog).[49]

In the catalog of the second exhibition (year IX), the actual purpose was already more precisely formulated: only perfectly executed objects should be displayed. Especially new inventions (*découvertes nouvelles*). As Bonaparte, who as consul opened the exhibition, clearly emphasized: "This solemn and memorable exhibition must calm all anxiety as to the future of our commerce." (Cf. *Seconde exposition publique des produits de l'industrie française* [Paris: Imprimerie de la république, year IX (1801)]).

Iron, transportation, industry interlock. Between 1840 and 1850 development by fits and starts is evident. England—in contrast to the continent—is hardly disturbed by political turbulence. Apart from this, England is in the forefront of industrial production in the first half century. Hence, the first Industrial Exhibition of All Nations took place in London in 1851.

[49]) One of the few surviving copies of the first exhibition catalogs is in the Conservatoire des Arts et Métiers in Paris, established in the same year—1798.

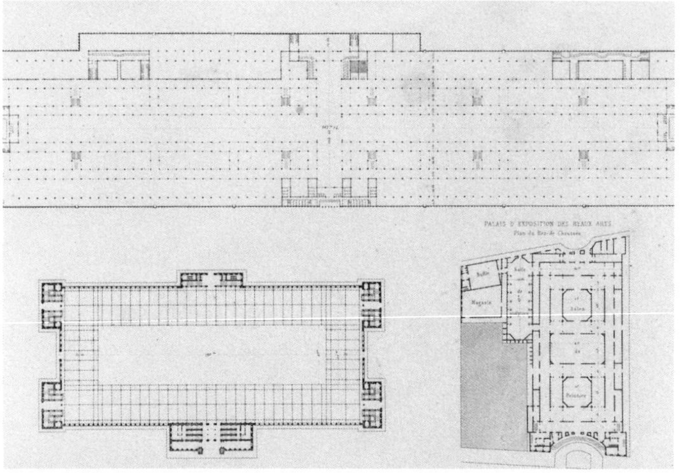

Fig. 28.
EXHIBITION, LONDON 1850.
Plan of the CRYSTAL PALACE
Material: cast iron. Span: 22 meters, thus less than late Gothic vaults. Innumerable cast-iron columns. Someone has counted 3,230. Fear of open space. But consistent in the pure structure of glass and iron.

EXHIBITION, PARIS 1855.
Main Building
The column-free space in the middle forms the nave with a 48-meter span.

Fig. 29. EXHIBITION, PARIS 1855. NAVE
Span 48 meters. Material: hand-wrought trusses. Cast-iron columns. First halls with such a bold span. Empire-style barrel vault. Even at this primitive stage of construction the meaning of the new material is sensed. Openness instead of spatial enclosure. Light. Lightness.

EXHIBITIONS OF WORLD COMMERCE

PARIS 1855

The first French event of this kind was called *Exposition universelle des produits de l'industrie*.[50] Compared with later exhibitions, both the public and the press still adopted a certain waiting attitude toward the exhibition of 1855. This is also sensed in the overall disposition: new participants[51] were constantly coming forward, so the

[50]) That is almost identical in name to the first show on the Champ de Mars, of 1798, only now, instead of the "French" limitation, the word was "*universelle.*" Global commerce.

[51]) Number of exhibitors: London 1851, 17,000; Paris 1855, 24,000; Paris 1867, 60,000; Paris 1878, 52,800; Paris 1889, 61,700.

Fig. 30. EXHIBITION, PARIS 1855.
Galerie des Machines
Gallery 1,200 meters long, attached to the main structure laterally along the Seine (cf. *L'Illustration* 1855).

Fig. 31. Pont de Longon (Garonne) 1854–55
Span of 77 meters. Height of girders 5.5 meters.

specified surface area proved to be too small, and the exhibition showed numerous additions. Plan: a rectangular structure with a high nave surrounded by a double row of galleries. The low, encircling aisles were supported by countless cast-iron columns. A circular panorama connected with the main structure and with a 1,200-meter-long gallery—containing mostly machines—that was oriented along the Seine.

Figs. 28, 29

Fig. 30

CONSTRUCTION: the soaring round arches of the nave seem much advanced, their span approaches a bold 50 meters. No tie-rods encroach upon the free space, yet one feels that the construction lacks a certain tautness that we are accustomed to today.
In fact, the hand of the constructor is here guided, both internally and externally, by historical concerns:

These neat, self-contained barrel vaults recall numerous palatial halls of the Empire period. And in order to resist the lateral thrusts of the large vaults, there was still no alternative but to imitate Gothic structural principles, that is, to buttress the iron, which was both expensive and wasteful of space.

Fig. 28 Compared to London's Crystal Palace of 1851, the span of whose central nave did not exceed 22 meters and thus remained below what had already been dared in the Gothic period—the 27-meter span of the wooden vaults in Padua's Il Salone—great progress is shown here. The architect L. A. Boileau, who incidentally in the same year erected the first church in Paris with cast-iron columns,[52] drew attention to the fact[53] that the Crystal Palace in London was only a large-scale reproduction of the glass houses in the Jardin des Plantes in Paris (erected 1833), which were consistently built of iron, and that in the first version of the Crystal Palace, which was dismantled shortly after the exhibition, the most difficult part of the construction, the dome, was even executed in wood.

The English, in fact, are less talented as constructors, but one must admit that they left the structure free, they did not encase it with a triumphal archway and stone walls as was done in Paris in 1855.

Cast iron was the construction material employed in London. In Paris in 1855 wrought iron was used for the first time for the vaults, although most pieces were still executed by hand.

Even without the appropriate structural solutions having been found, in this exhibition attempts had already been made: large spans, unusual luminosity, lightness of execution.

PARIS 1867

PLAN: two semicircles of identical diameters joined by two straight lines form the
Fig. 32 outer contour. Small axis 380 meters, large axis 490 meters. (The site of the Champ de Mars, between the Seine and the Ecole Militaire, was selected for the first time. The oblong form of the site defined the outer contour for this and subsequent exhibitions.) The question arose of how one should dispose the individual sections so as to give the spectator an ordered and undisturbed insight into the whole development.

The elliptical ground plan was filled with seven concentrically arranged galleries; the
Figs. 34, 35 garden was laid out inside the innermost ellipse of this *colisée du travail*.

Proceeding outward, the galleries grew progressively wider and larger. The outermost gallery, the Galerie des Machines, with twice the width and height of the others, towered over them all. Behind this belt followed encircling rings: clothing, furniture, raw materials, until the two innermost—smallest—galleries, in which a retrospective exhibition, *l'histoire du travail,*[54] and the fine arts were displayed. A palm garden with sculptures formed the open, innermost oval. The oval building was

[52]) Saint-Eugène. Though Boileau resorted to imitating the Gothic columns of a refectory (Biblioth. du Conservatoire des Arts et Métiers).

[53]) This is not to be taken literally, for John Paxton used his own greenhouses as a model in his Crystal Palace, or, as the French called it, in his "*cage de fer.*" (L. A. Boileau, *Le fer, principal élément constructif* DE LA NOUVELLE ARCHITECTURE [Paris, 1871], p. 71). In any case, it should be mentioned that Paxton used wooden ribs in his large greenhouse for the duke of Devonshire in Chatsworth (1837–41).

[54]) The exhibition also concerned itself with social problems. Workers' housing was built, and "the Universal Exhibition of 1867 prompted new inquiries into the state of the working classes." It presented

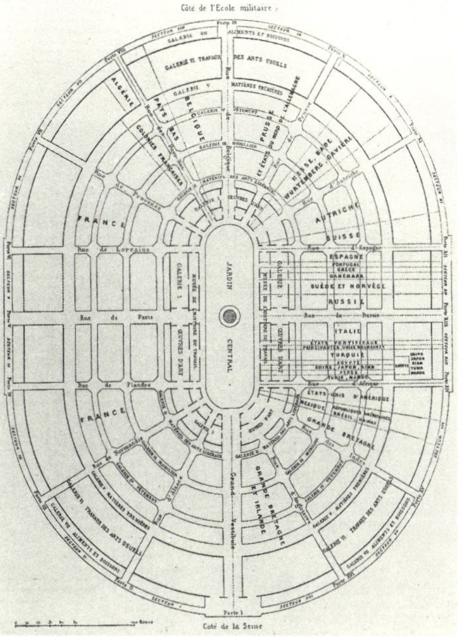

Fig. 32. EXHIBITION, PARIS 1867.
PLAN OF THE MAIN BUILDING
(On the Champ de Mars between the Seine and the Ecole Militaire.)
Attempt to line up the products of different countries and parts of the world next to each other, almost as a SET OF STATISTICS. Arranged peripherally: the EXHIBITS. Radial avenues connect the seven rings to permit visitors to examine the total production of a country at a glance.

divided into individual segments by transverse avenues, so that the spectator could Fig. 35
follow effortlessly the development of a single country by itself and in relation to others. Attempt at a set of "living" statistics. In reality the system did not work due to the unequal development of the various countries.
"To go round this palace, circular, like the Equator, is literally to go around the world. All peoples are here: enemies live in peace side by side. As with the beginning of things on the watery globe, the divine spirit now floats on this globe of iron." (Cf. *L'exposition universelle de 1867 illustrée: Publication internationale autorisée par la commission impériale,* vol. 2, p. 322).

CONSTRUCTION: the iron skeleton of the Galerie des Machines (span of 35 meters, height of 25 meters) consisted of pillars reaching 25 meters high. They received the basket arches of the vault girders at the height of 19 meters. These pillars were allowed to continue past the spring of the vault straight up into the air.[55] The chief constructor, J[ean-] B[aptiste-] Séb[astien] Krantz, did not want to expose tie-rods within, and the lateral thrust could only be borne by free girders that punched through to the sky from the pillars to the basket arches. These projecting pillars were considered hideous, and one attempted to disguise them with trophies and flagpoles.
Hydraulic lifts, with all the clumsiness of newborn monsters, afforded access to the Fig. 33 [36]
peak of the roof, around which ran a platform allowing an unusual view into this
gallery city of corrugated sheet metal and glass. Fig. 34
Behind Krantz, the chief, stood the actual inspirator, the young Eiffel, who

"objects specially exhibited with a view to improving the physical and moral condition of the people." Cf. G[eorges] Weill, *Histoire du mouvement social en France, 1852–1924* (Paris, 1924), p. 89.
[55]) A similar solution on a more modest scale was attempted in the Palm Garden in Frankfurt a. M. of 1871.

Fig. 33. Ed[ouard] MANET, EXHIBITION, 1867
MANET not only grasped externally the reality of the age but also translated it into an adequate pictorial language: colors dissolve the fixed form, the verifiability of load and support; what counts in this image is the suspended relation between figurations of color.
(Suggested in the distance at the right is the exhibition building with its soaring piers.)

Fig. 34. VIEW FROM THE PLATFORM
of the seven-ring gallery of glass and corrugated sheet metal.

Fig. 35. EXHIBITION 1867. JARDIN CENTRAL
with sculptures in the innermost oval and an encircling gallery of slender cast-iron columns.

had recently established his firm in Levallois-Perret. Eiffel calculated and later experimentally verified the large supports of the Galerie des Machines, which were the core of the entire arrangement. Thus for the first time the coefficient of elasticity for a large-span building was verified. Eiffel later published his findings in a memoir.
These tall, kilometer-long galleries were undoubtedly grand. They were filled with the roar of machines. It should not be forgotten that people still arrived at the particularly famous festivities of this exhibition in eight-horse carriages. As with contemporary rooms, one attempted to reduce the scale of these 25-meter-tall galleries with furniture-like installations and to soften the severity of the construction. One was afraid of one's own greatness.

PARIS 1878

The success of 1867 was so great that the scale of the next exhibition was significantly enlarged. It was divided into two sections. A monumental structure and a temporary exhibition building. Across the Seine in 1876, Davioud and [Jules-Désiré] Bourdais built the Trocadéro as a masonry palace.[56]
The actual exhibition structure was again defined by the elongated form of the Champ de Mars. A rectangle: 350 × 700 meters. The earlier elliptical form was

[56]) These monumental buildings inherently contradict the meaning of an exhibition, and they led—by 1900—to the collapse of the exhibition idea. The Trocadéro is cited only as a symptom. Davioud and Bourdais were excellent architects and are to be thanked among other things for their design for a National Opera for an audience of 5,000, which for the first time used acoustics as a design principle.

Fig. 36.
EXHIBITION, PARIS 1867.
GALERIE DES MACHINES
Mechanical elevator to the roof platform.

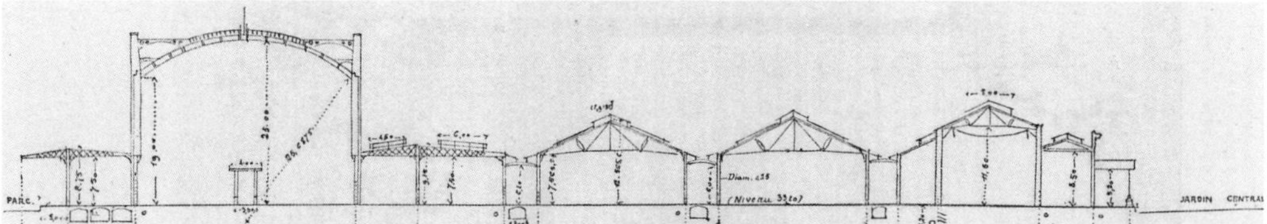

Fig. 37. SECTION THROUGH EXHIBITION BUILDING 1867
In order to avoid visible tie-rods, KRANTZ and EIFFEL extended the pillars of the superelevated GALERIE DES MACHINES into the sky to deflect lateral thrust externally.

Fig. 38. GALERIE DES MACHINES 1867
In the center of the gallery ran a platform on cast-iron columns, which served as a carrier for the transmissions of the machines.

Fig. 39. ENCIRCLING OUTER GALLERY 1867
It included restaurants from each country. CANOPY that would soon be replaced with glass. [François] COIGNET used CONCRETE in the basements for the first time.

Fig. 40.[57] First thoroughly consistent **IRON-SKELETON BUILDING. 1871–72**
Menier Chocolate Factory in Noisiel-sur-Marne near Paris. Architect: Jules Saulnier
Cantilevered on four piers, the building is suspended above the water on box beams.
Instead of a rough surface of advancing and retreating elements, the architect, guided by the construction, astonishes us with a thoroughly planar facade. As the architect himself expressed it: "The system of construction adopted in the facades yielded an entirely flat surface from top to bottom, without any horizontal or vertical projection" (cf. *Encycl. d'arch.*, 1877, p. 92).
We feel: the "planar surfaces" of the "new" architecture have their origin far in the past.
Iron breaks through to the surface. The diagonal braces still show clear reliance on half-timbering. But the primitive state of the construction protects it from being draped with false architectural hangings.

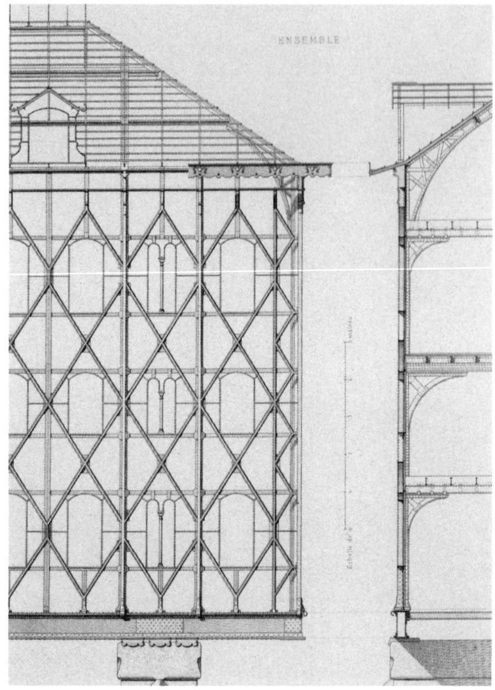

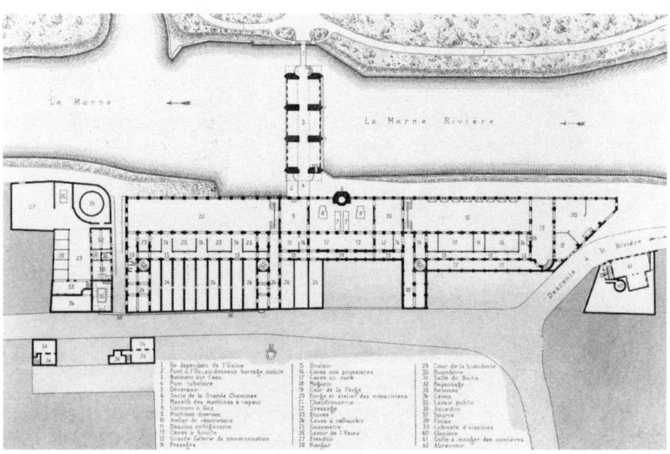

Fig. 41. JULES SAULNIER
Menier Chocolate Factory in Noisiel-sur-Marne
Plan. The main building on the riverbank dates from 1864–67. The iron-skeleton building shown above is the one that rests on four piers and is set boltlike into the riverbed.

57) To suggest the evolutionary path, we are throwing in a few buildings in the "Exhibitions" chapter. We should further add here examples that we discussed in "FIRST FORMATION."

Fig. 42. The IRON SKELETON as HOUSING FORM. 1927. Apartment block of the Stuttgart Werkbundsiedlung by Mies van der Rohe. 1927
It took more than half a century before the importance of the iron skeleton for apartment houses was recognized. The conclusion to be drawn here from the construction is: fixed interior walls are senseless in this type of construction! Each tenant should be given the opportunity to arrange his dividing walls freely according to his own needs. Industry is given the task of constructing such walls flawlessly.

Fig. 43. PONT DU BELLON. 1868–71
Two pylons. Height 48 meters. Total length: 231 meters. Bridge girder: 45 meters.
From the beginning, the French used slender iron frameworks instead of stone pillars (cf. [Edouard] Collignon, *Chemins de fer;* contains photos of bridges, railroad stations, etc., from the period 1850–1880).

Fig. 44. EXHIBITION, PARIS 1878. VESTIBULE, MAIN ENTRANCE Facing the Seine.
CONSTRUCTOR: EIFFEL
The representational inflated sheet-metal architecture of the domes is only a part of the entire picture. What is important here is the courage with which the glass and iron are functionally united: the GLASS WALL and the unity with which the glazed canopy projects. These canopies significantly undercut the viewer's preconceptions concerning the relations of load and support in a structural system.

replaced by a quadrangle, for one had learned that the curved girders of the inner supports were more difficult to reuse than straight ones (*Encycl. d'arch.*, 1878, p. 36).

A series of galleries in parallel formation extended over the length of the site; all were flanked and dominated by the two Galeries des Machines. Vestibules were placed on the two short sides of the site in front of the galleries. Eiffel constructed
Fig. 44 the monumental vestibule facing the Seine. The corners and middle of the vestibule were accentuated by great *dômes métalliques,* whose inflated volumes and dubious architecture were described already by contemporaries as "highly questionable" (*Encycl. d'arch.*, 1878, p. 62). In these domes, iron was violently forced to assume monumental forms alien to the material. The result is obviously an inflated sheet-metal architecture.

Nevertheless: it should not be forgotten that along the length of the entire vestibule ran a glazed canopy, a *marquise vitrée.* Unpretentious and perfectly light. In these details one detects the hand of Eiffel, who at the same time (together with L. C. Boileau) placed a similar *marquise vitrée* on the first iron-skeleton department store (Bon Marché).

If we were to scrape the decorative sludge off these buildings and make it a habit impartially and urgently to inquire into

→ 1926

Fig. 45. W[alter] GROPIUS: BAUHAUS DESSAU 1926
Only after half a century are we able to explore the tensions[58] inherent in the materials and overcome the decorative incrustation.

58) Tension in the aesthetic sense.

1926

Fig. 46. J[an] F[rederick] STAAL: SHOPPING STREET (Jan Evertsenstraat), AMSTERDAM 1926
The freely cantilevered canopy is only now beginning to be used unaffectedly.

their true nature, we would see that their bodies already contain all the essential building elements that we today describe as new. We could compare the glass wall of 1878 with the glass wall of the Bauhaus just as easily as with the Glass Building department store in San Francisco, 1878, or we could compare the running glazed canopy of 1878 with the canopy of Fig. 46 the shopping street in Amsterdam of 1926, where it becomes a truly articulating element.

Fig. 47 CONSTRUCTION. The essential point: the two flanking Galeries des Machines. 35-meter span, 25 meters high.

The barrel vault has disappeared. The form resembles a capsized ship's hull. The truss girders meeting at the ridge of the roof are distinct and separate elements and already show that this is not a rigidly continuous construction.

Pier height 16 meters. They are box-shaped with a quadrangular cross-section. The truss girders begin halfway up.

SUPPORTS: de Dion type. The engineer Henri de Dion was the real creator and pioneer of girders for very large spans. After the most careful studies of material tensile strength he arrived at the form truss girders should have in order to be able to withstand, without the aid of tie-rods, the various stresses placed on them. De Dion died shortly before the exhibition opened, while still working on his calculations. The profile of the truss girders already shows a certain inner elasticity, the result of studying the actual laws of the materials.

Stepped purlins running through and connecting the lattice girders gave the girders in their continuity an expression of precise repose such as had not previously been realized.

On both sides, from the halfway point up, the walls were filled in with glass. Such a union of glass and iron, by its nature, demands an extensive dematerialization of the building, as can already be seen here. Boileau precisely describes the expression produced by this union: "The spectator is not aware of the weight of transparent surfaces. The surfaces are to him air and light, that is to say, imponderable fluidity."[59]

Because of the Dion truss it has become possible to transmit all apparent forces directly into the foundation without outside help. Nevertheless, the foundation remains rigidly connected with pillars and framework: the pillar is riveted into U-shaped iron sockets that are sunk into the foundation. But an iron skeleton is subject to changes, it cannot be rigidly bound together like a stone palace. It lives with the temperature fluctuations.

One begins to take this into account: each 60 meters along the ridge of the roof, where the two girders meet, a complicated system of bolts and oval holes permits independent expansion and contraction of the entire skeleton.

Perhaps there is still in the rigid connection with the ground, in the box-shaped girders of the lower part, a memory of the old relationship between column and base. But after the—if you will—capital-like enlargement of these girders, at the mounting of

[59]) "*Les toitures et les plafonds vitrés ne supportent pas une construction d'apparence massive ou compliquée*" [Glazed roofs and ceilings are incompatible with any construction that looks massive or complicated]. "*Le spectateur n'admet pas la pesanteur des surfaces transparents. Pour lui ces surfaces représentent de l'air et de la lumière, c'est-à-dire des fluides impondérables.*" *Encycl. d'arch.*, 1887–88, p. 97.

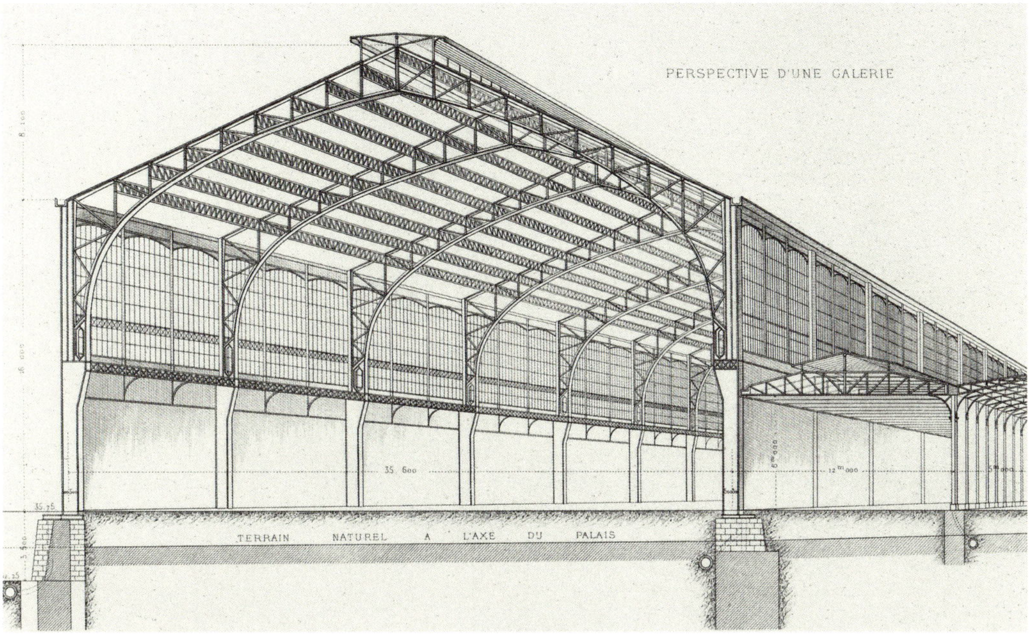

Fig. 47. EXHIBITION, PARIS 1878. GALERIE DES MACHINES. CONSTRUCTOR: HENRI DE DION
Span 35 meters. Height 25 meters. For the first time in a large span all the forces connected with the system are carried freely into the foundation without auxiliary means (tie-rod).

the framework, there is a new tensile elasticity. One begins to recognize that the expression of iron skeleton also represents something new, demanding a hovering balance of forces.

PARIS 1889

The exhibition of 1889 is both the climax and—interpreted from the standpoint of knowledge—the conclusion of this development. Its influence was enormous.
Plan: dispersed layout. The Tour Eiffel by the Seine was the focal point, behind which the court-shaped exhibition buildings receded. The two wings contained the *beaux-arts* and the *arts libéraux;* a section devoted to general exhibitions connected them. The Galerie des Machines with its immense metallic nave rose in the background to tower over the complex.
CONSTRUCTION: GALERIE DES MACHINES. (Actually called the Palais des Machines.)[60] Constructor: Cottancin.[7] Architect: [Charles-Louis-Ferdinand] Dutert. The dimensions exceeded anything previously known. The largest previous internal span of 73 meters wide and 25 meters high had been tried in England in 1868.[61] Fig. 48

[60]) In 1910 this culmination of constructional design was torn down—as Frantz Jourdain correctly notes—out of "artistic sadism." This hall was imitated in the Chicago World Exhibition of 1893 where it served as the palace of art, costumed as a Palladian basilica and senselessly imprisoned behind walls.

[61]) At London's Saint Pancras Station, though of a similar form—also without columns—the girder framework is rigidly connected to the ground with heavy anchors. The lateral thrust is offset by tie-rods buried beneath the ground. Constructor: [William Henry] Barlow. Fig. 49

Fig. 48. GALERIE DES MACHINES, PARIS 1889
Assembling of the hall. One sees clearly how the two trusses are connected only by a bolt at the ridge line. Instead of a rigid connection, some movement is allowed at the top. Joint.

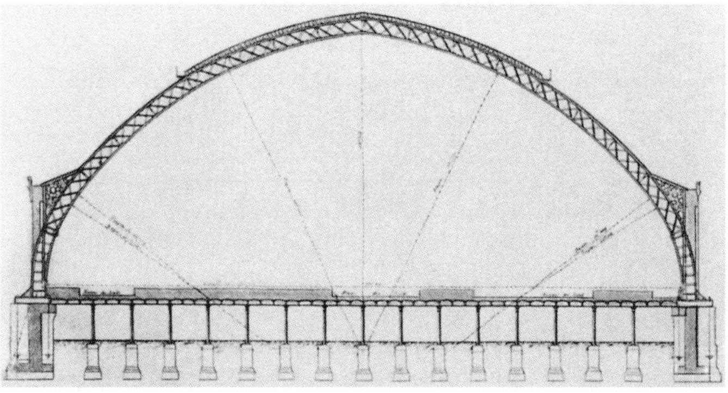

Fig. 49. SAINT PANCRAS STATION, LONDON 1868
Span 73 meters, internal height 25 meters.

The Palais des Machines spanned 115 meters with a height of 45 meters, that is, it exceeded the height of the nave of the cathedral of Amiens, whose span was about one-eighth (15 meters) of this hall. The skeleton is formed of twenty trusses. Total length 420 meters. Huge glass walls enclosed the sides.[62]
The freely spanned spatial volume signified an unprecedented conquest of matter. Nothing from an earlier time can compare with it.
The glass walls at the end did not close up the structure. This is not a building that rests within itself. There could be more trusses, there could be fewer trusses lined up, which would not decisively alter the building, for the aesthetic meaning of this hall lies in the perceptible union and in the penetration of the

[62]) "The large truss has a grand and bold profile—the forms of the metal were everywhere so devised that it constituted its own decoration" (EXPOSITION UNIVERSELLE *de* PARIS *de 1889, Monographie. Palais, jardins, constructions diverses et installations générales,* par A[dolphe] Alphand, 1892–95), p. 521. On a smaller scale, J. W. SCHWEDLER had previously attempted the free "pointlike" support in what was from the point of view of construction and aesthetics the most beautiful hall of the age: the Central Station, FRANKFURT A. M., dedicated 1888. The tripartite halls of the Frankfurt railroad station indeed have trusses that touch almost in points at the apex, but which do not lose the barrel-vault form that the TRIPLE-HINGED GIRDER of the Galerie des Machines overcomes intrinsically through its essence. Dimensions of FRANKFURT: 186 meters LONG, 56 meters WIDE, 29 meters HIGH.

Fig. 50. GALERIE DES MACHINES, PARIS 1889
Upper gallery. View into the spacious hall.

Fig. 51. GALERIE DES MACHINES, PARIS 1889
"Baseless," the weight of the entire hall rests on small hinged joints, without rigid connection to the ground.

exterior space, resulting in an utterly new limitlessness and movement, in keeping with the rotation of the machines that filled it. Each arch was made up of two segments. They were joined—hinged—with a bolt at the apex of the hall.
Below, the girder gradually tapers so as barely to touch the ground.
Above, it expands, achieving enormous impact. The usual proportions seem to be exactly reversed, the traditional static feelings disturbed and disrupted: TRIPLE-HINGED ARCHES.
Further proportional distensions:[63] the trusses are approximately five times as high as they are wide (3.5 meters high by 75 centimeters wide).[64]
With these truss proportions in an enclosed space we had to become aware that—in contrast to stone or wood construction—the filling material was missing. These trusses were unusually light because this was the first time that a steel framework was used to such an extent. The eye of the contemporary onlookers felt insecure and disturbed as the light pouring in from above swallowed up the thin latticework. The vault optically attained an unusual hovering state.[65]
The last hint of columns has disappeared, it is impossible to discern where support and load flow into one another.
The arching begins quite low,[66] bent as though in the act of leaping in order to receive the load. If you will: this is the symbol of our caryatids: they bear their load with neither the dignity of antiquity nor the buckling of the Baroque. They spring toward the load in order to unite with it.
The ends of the downward-tapering girders are no longer rigidly connected with the ground but are left free to move.[67] They transmit their weight, as well as a horizontal thrust of 120,000 kilograms, directly into the foundation by a hinged joint. With this supporting structure even foundation movements can take place without creating internal stresses. This was the only means of controlling the play of forces at all points.[68]
The division of support and load, which was still suggested in de Dion's halls of 1878, is here obliterated.

63) A[rthur] Vierendeel, *L'architecture métallique au XIXe siècle et l'Exposition de 1889, à Paris* (Brussels, 1890), p. 30: "This disproportion makes a very bad effect, the beam is not balanced, it is not firmly seated, the eye is not reassured."

64) This was seen as a distortion, for the eye demanded as an eternal point of comparison the dimensions of the stone architraves.

65) Critique of contemporaries: Vierendeel, p. 31: "The keystones of the Palais des Machines have yet another fault, which is that they are too much hollowed out."

66) Vierendeel, p. 29: "The curvature of the vaults is also highly defective from an aesthetic point of view. It begins too low." Precisely these criticized points were retained for later development. Contemporaries, even a de Baudot, found the "proportion" wanting!

67) As often happens in the nineteenth century, important results come about when various fields overlap: one took over the principle of free support from bridge construction. At the same time one sees in such instances how difficult it is for human knowledge to carry over the results of one field to another.

68) "Only one mode of construction provided a mathematical determination of the distribution of thrusts in the different parts of the curve: this was that of articulation at the support and at the crown. This system guarantees an absolutely rigorous rational distribution of the thrusts and of the materials." Cf. A. Alphand, *Monographie 1889*, vol. 1, p. 46.

Fig. 52. GALERIE DES MACHINES, PARIS 1889
View toward the main entrance. DIMENSIONS: 115 meters INTERNAL WIDTH, 45 meters HIGH, 420 meters LONG.

Fig. 53. GALERIE DES MACHINES, PARIS 1889
MAIN ENTRANCE

Fig. 54. TONY GARNIER: SLAUGHTERHOUSE IN LYONS
Dimensions: WIDTH 80 meters, LENGTH 210 meters.

Fig. 55. BRIDGE OVER THE FIRTH OF FORTH, 1883–89
CONSTRUCTORS:
JOHN FOWLER and [Sir] BENJ[amin] BACKER

Three gigantic steel pylons 100 meters high, span of each superstructure 520 meters. It employed nearly 5,000 workers. (Cf. *Encycl. d'architecture,* 1888/89, pp. 166 and 186.)

The iron skeleton has found its true form.
A play of enormous forces is held in equilibrium. But not rigidly, like support and load, rather, almost floating.
It is the equilibrium of a balance beam daringly poised against continually varying forces.
A new oscillating harmony is created.
An elastic counterpoise is achieved with respect to changes within, without, and in the foundation:
Equilibrium with respect to change in its own molecular structure.
Equilibrium with respect to external pressure (wind, snow).
Equalization with respect to the surface fluctuations (foundation).

CONSTRUCTION BECOMES EXPRESSION.

CONSTRUCTION BECOMES FORM.

**Fig. 56. EIFFEL TOWER 1889.
A PIER. EXTERIOR**
From the ground floor to the first platform.

FURTHER DEVELOPMENT

The significance of the Eiffel Tower lies in its structure. By its silhouette undoubtedly a product of its age: monument, sculpture. But all flesh has been left off, everything is reduced to connective parts, and the air drawn into the interior of the piers now becomes, in an unprecedented way, a formative material. Eiffel and his engineers erected the tower in seventeen months. Each rivet hole had been factory drilled in advance to a tolerance of one-tenth of a millimeter, whereas the English did this work on-site for the bridge over the Firth of Forth (1883–89). **Figs. 2, 56 to 58** **Fig. 55**

Fig. 57. EIFFEL TOWER 1889
View from the second platform to the first. Taken from inside the elevator shaft. Right and left the curved elevator rails that continue down to the ground floor.

One might also view the Eiffel Tower as a manifesto: the successful realization of a 300-meter-high structure provided the decisive impetus for tall steel-skeleton buildings.

This development continued underground: America. The question of who built the first skyscraper is still being disputed. We know only that in Chicago—apparently at the same time as the Eiffel Tower—quite a number of architects started building the first eleven- or twelve-story steel-framed houses.[69]

In Paris itself—quite apart from department stores—a series of modest skeleton structures with iron facades was erected in the following decade.

An entire square was even placed on an iron foundation. It was the place de l'Europe, the generous structure over the wide network of rail lines from the Saint-Lazare Station. (Engineer: Julien). With a half-dozen intersecting streets.

The tradition of the nineteenth century is today being renewed. Many efforts in

[69]) "Who Designed the First Steel Skyscraper?," *The Western Architect* 32 (November 1923), p. 125. It is mentioned that the architects were [John] Root, [Dankmar] Adler and [Louis] Sullivan, [William Le Baron] Jenney, [William] Holabird, and [Martin] Roche in Chicago as well as L. [Cass] Gilbert (eleven-story Tower Building) in New York.

Fig. 58. SUSPENDED STAIRS within the EIFFEL TOWER
(Connection from the ground floor to the first platform.)
One sees the elevator track next to the stairs. Our modern intentions find precedents even in the detailing of forms such as the horizontal railings of the airy staircase. Cf. fig. 62.

1889

1926

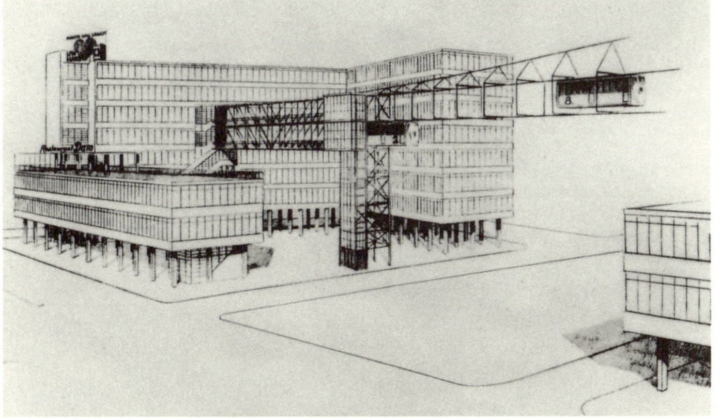

Fig. 59. MART STAM: SUPERSTRUCTURE OF THE ROKIN DAM AMSTERDAM 1926
Only now do the seeds that lie in structures such as the Eiffel Tower come to full fruition. The affinity with a building such as the Eiffel Tower lies not merely in the connection and interpenetration by suspended transportation or free-hanging stations; one reaches the conclusion viewing both buildings: ARCHITECTURE NO LONGER HAS RIGID BOUNDARIES.

Fig. 60. PONT TRANSBORDEUR, MARSEILLES. CONSTRUCTOR: [Ferdinand-Joseph] ARNODIN. 1905
The ferry is suspended at a height of 51 meters from a trolley that runs on a gangway. The FERRY glides a few meters above the water. Length of gangway 240 meters. Graceful combination of stationary and moving parts.

Fig. 61. PONT TRANSBORDEUR 1905
View from the stairs toward ferry and city. The oldest "suspended ferry" of this kind is in Rouen—also executed by ARNODIN, in 1899—yet one finds others, for instance, in Nantes, and the largest, still under construction, in BORDEAUX.

Fig. 62. PONT TRANSBORDEUR 1905

View from the elevated gangway, 54 meters above the water, to the suspended ferry. Plenty of new visual possibilities: everything is based on mobility. Notice the platforms of the staircase that boldly project out into space. The "new architecture" has unconsciously used these projecting "balconies" again and again. Why? Because there exists the need to live in buildings that strive to overcome the old sense of equilibrium that was based only on fortresslike incarceration.

Fig. 63. View of Houses from *Pont Transbordeur* Marseilles

Fig. 64. VELODROME d'Hiver, Paris

England, France (through Henri SAUVAGE), and Germany are being made to use the SKELETON building even for HOUSING SETTLEMENTS. Research and experience in this area are still very much in the initial stages. It is not yet possible to determine beyond doubt their economic feasibility.
The historian must therefore establish that by midcentury the problem of mountable and transportable iron housing with cavity walls was already posed. Such houses were built for the tropics (colonies) from the perspective: thermal insulation and resistance to insect attacks (termites).

Fig. 65. VELODROME d'Hiver, Paris
Gallery bleachers.
The bleacher rows that are perforated and cut in order to achieve lightness enable new effects through light, structure, material variety.

FERROCONCRETE

It is pointless to discuss the new architecture in France without touching upon its foundation: ferroconcrete. It is not extracted from nature as a compact material. Its meaning is: artificial composition. Its origin: the laboratory. From slender iron rods, cement, sand, and gravel, from an "aggregate body," vast building complexes can suddenly crystallize into a single stone monolith that like no previously known natural material is able to resist fire and a maximum load. This is accomplished because the laboratory intelligently exploits the properties of these almost worthless materials and through their combination increases their separate capacities many times over. We know: a load-bearing beam—be it a bridge girder or ceiling joist—is chiefly subjected to compression in its upper part and to tension in its lower part. Therefore, iron, which possesses excellent tensile strength, is placed more on the underside, whereas concrete, with its great compressive strength and compact mass, predominates in the upper part.

[Joseph] Monier did not know this—in 1867. In his reinforced-concrete containers iron gave the form and concrete was the filling. Being more persistent than his predecessors and contemporaries—[Joseph-Louis] Lambot (1854), Coignet (1861),[70] [Thaddeus] Hyatt (1877)—Monier elaborated his system step-by-step and successively took out patents for pipes, flat slabs, bridges, and staircases (1875). Despite an instinctively correct arrangement he failed even at the end to recognize the function of iron and concrete. This insight fell to German engineers in 1880. But the decisive step that would enable a new means of architectural design to arise from an ancillary material, from a construction detail, was taken by Fr[ançois] Hennebique.[71]

Even this, like almost all enduring knowledge of our age, was not the result of fantastic visions—they were later consequences—but of microscopic examination. Or, expressed in the more modest language of the our time—of patents.

In 1892 Hennebique took out a patent on "composite beams."

Until then the weak point in ferroconcrete construction had been the joints: those places where the ceiling merged with the beam and the beam with the supports. It didn't work there. With an appropriate placement and bending of the reinforcing iron, Hennebique succeeded in bonding the ceiling, beams, and columns into one

[70]) To our knowledge, the first extensive use of CONCRETE occurred at the Paris Exhibition of 1867. In the large ellipsoidal main building (cf. fig. 39) Coignet used concrete ceiling slabs above the cellars of the commercial areas that were located in front of the Galerie des Machines.

[71]) François HENNEBIQUE, who came to ferroconcrete via stone masonry and iron construction, is himself one of the first examples of an industrial building entrepreneur. Just as one had previously sold machines to every part of the world, he erected his constructions in the most widely different countries: mills in France, granaries in Genoa, silos in Strasbourg with the help of Wayß & Freytag, harbor structures in England, the first wide-span bridge over the Rhone (near Vienne); the decorative confections of the Paris Exhibition palaces of 1900 likewise conceal Hennebique's path-breaking concrete frames.

continuous unit (ribbed beam). The building as monolith became possible. Isolated cast-iron columns were replaced by ferroconcrete pillars. The unified frame-structure emerged. The architectural imagination could move in tow of the patents. We know how cumbersome this process was.
It is clear where this path would lead. Formerly little-noticed uprights were now busily examined. In 1899 Armand Considère discovered special methods to make them rigid (corded concrete). Most recently one has even succeeded in making the beams—these last vestiges of timber construction—disappear: girderless ceilings! The efficacy of iron radiates almost like a magnetic field over its corporeal expanse within the concrete (developed by the Swiss [Robert] Maillart), and to be sure in a more empirical than mathematically acceptable way. We are faced with a very complicated interplay of forces that even our theory can scarcely calculate.
Meaning of ferroconcrete: this laboratory product, this product that emerged only by focusing on the material, through—if you will—a material vision, is very significant for the present and the future. With its prevalence, the architect, as a romantically sketching hero, became an embarrassment. No material avenged its violation through its obstinate behavior as much as did ferroconcrete! On its framework one could, it is true, attach the most outrageous facades, but its actual control, despite possible exterior mutilation, belonged to the engineer. And behind the engineer: industrialized building production.
The concrete firm is not the conventional building entrepreneur who passes the work on to about fifty manual laborers and actually does nothing more than continue the medieval building production in a coarser way. Scientific, industrialized building production stands behind the concrete entrepreneur. Ferroconcrete requires this, from its laboratory beginnings to the treatment demanded by the material on-site. There is no other way. Here we point out its characteristics; this is not the place to expand on the many shortcomings ferroconcrete still possesses today that impede its applicability. It will hopefully not remain the only new material. To say nothing of the fact that wood and iron also are about to be rediscovered.
What is important: to show that we find ourselves in a lawful evolution and that our real tradition has gradually been formed during the last half century. In its strength and depth it is comparable to any other age. Only we must place the accents correctly.
The discovery of Monier-Hennebique-Considère was initially passed to French architects. In 1903 the brilliant constructor Perret already drew the practical conclusions in his house on the rue Franklin. The banks declined him a mortgage because the Fig. 68
experts predicted that the slender frame would collapse.

NATIONAL CONSTANTS

Fig. 66. LIMOUSIN AND COMPANY (Freyssinet Technique). ORLY HANGAR 1916
View of one vault ELEMENT in a finished state and of another with scaffolding. Each element has a span of 80 meters, width 7.5 meters, height 56 meters. One arch was erected every eight days.

We want an international architecture. An architecture for the age. All living nations are moving toward it. Nevertheless, each country has its own predetermined role in the movement. This is already clear today. Precisely the new avant-garde architecture of France, which has rapidly freed itself from traditional formal games, is a sign that the native primeval voices speak again. Naturally, not in imitation of the royal styles. Even in France the new architecture is being reproached: it is international! Hopefully it is that! Likewise we cannot overlook that in its entire method, in how the tradition of iron and ferroconcrete was founded and amazingly developed, the threads lead backward: Gothic. The same soil produced the cross-ribbed vault. The same urge to lighten matter, to demand of stone what apparently goes beyond the strength of stone, has returned. The strands lead from the almost fragile, slender arches in the choir of Beauvais to the enormous concrete parabolas of the Orly hangars built south of Paris (1916) by the engineer [Eugène] Freyssinet, commissioned by the firm of Limousin & Co.

Once again we should add that the battlefront between national and international no longer exists in reality, just as today an actual battlefront is basically no longer drawn between states, but in the ongoing struggle about SOCIOLOGICAL STRUCTURE!

The state of this problem differs from one country to the next, but the problem itself is everywhere the same. Architecture is as closely bound to the sociological structure of a country as to its climate, materials, customs. A fruitful variety of types naturally evolves from a common ground.

When we describe lines arching back in time, we are doing so in order to emphasize the French constructional temperament.

The constructional temperament of FRANCE is as indispensable for the new architecture as America's organizational aptitude or—in its place—Holland's handicraft aptitude. Each country must in its own way contribute to collective advancement.

1900—1920

In the nineteenth century the struggle between the functional architecture of rationalism and academicism always ended with the academy winning. The particular time was simply not yet ripe, either in its means or in its knowledge, to prevail. It cannot be denied: the past proved itself to be stronger. Only today can the past be finally put aside, for a new way of living [*Lebensform*] demands a breakthrough. This new way of living is to a large degree equivalent to the expression anticipated by, and latent within, the constructions of the nineteenth century.

The last struggle between functional architecture and academicism is seen in the generation whose pathbreaking works occurred shortly after 1900. Auguste and Gustave Perret (Paris)—born 1874 and 1876—and Tony Garnier (Lyons)—born 1869—generation-mates of the pioneering wave: Lloyd Wright (born 1869), Adolf Loos (born 1870), van de Velde (1863), Berlage (1856). Both Perret and Garnier are students of the academy. Garnier even a prizewinner and *pensionnaire* of the Villa Medici. Both carry the academy within. This is their limitation. It has been claimed that the school did not harm them, but in reality they are directly bound in their design to the classical French ideal. It surfaces at every opportunity and roams through their buildings. Disguised and undisguised.

Fig. 67. Aug. Perret on the Roof Terrace of His House, 25 rue Franklin
Terrace and railings (made of automobile pipes) built in 1903. Picture taken in 1927.

Fig. 68. A. G. PERRET: HOUSE, 25 bis RUE FRANKLIN, Paris
First undisguised skeleton-frame building in FERROCONCRETE. Slender ground-floor dimensions, hollowing out of the compact facade, soaring dissolution of the top of the building.

A. G. PERRET

Perret is a constructor.[8] Engineer-architect. From the beginning. He comes from a Burgundian family near Cluny. One might draw a connecting line,[72] if one were so inclined, between the severe art of engineering construction of the Cluniac monks and the architect who, for the first time, understood how to translate ferroconcrete into architectural expression.
In his apartment house in Paris, 25 bis rue Franklin, 1903, ferroconcrete is used for the first time for a RESIDENTIAL BUILDING, and the facade openly shows the skeleton frame as a constituent element.[73]
In this narrow apartment building reside, almost like a vision, the germs of later development that Le Corbusier and others elaborated: the planar facade is shattered. It is hollowed out, recedes in depth, springs forward again: allowing six cantilevered stories to jut out, freely suspended, with the sixth story exposing naked, rectangular piers. The whole facade is in movement. The roof already carries the rudiments of a garden. Above, the building is almost suspended, and on the ground floor—where one notices the thin concrete ceiling slabs of the shops—there are only slender concrete piers, otherwise no mass remains. The building also grows lighter at the base, it approaches the iron constructions that touch the ground only at points.
No light well. In the narrowest space a staircase brightly lit by daylight, as glass bricks were used for the walls, a material

[72]) Cf. Paul Jamot, *A. G. Perret et l'architecture du béton armé* (Paris and Brussels: Edition G. Vanoest, 1927)
[73]) Hennebique's residence on the rue Danton uses ferroconcrete only as a concealed substitution. The ceilings of the first Perret building, the Casino at Saint-Malo of 1899, were also of ferroconcrete, but it should not be forgotten that many instances of its use in utility buildings preceded that.

Fig. 69. A. G. PERRET: RUE FRANKLIN, 1903
View from the uppermost roof terrace to the eighth story. The elements of later loosening are already clearly present.

that, as is well known, only much later found acceptance in residential construction.[74] The next step is the garage on the rue Ponthieu (Paris) 1905. The masonry **Fig. 72** fill is dropped. A thin structural framework and glass surfaces predominate; in a

[74]) Perret explained to us that he used these glass-brick walls because a neighbor could legally have forbidden him from installing windows on this side of the building. This detail nevertheless naturally belongs in the series of progressive applications of materials, just as do the railings of exhaust pipes on the uppermost roof terrace or the dissolution of the ground floor into glass walls.

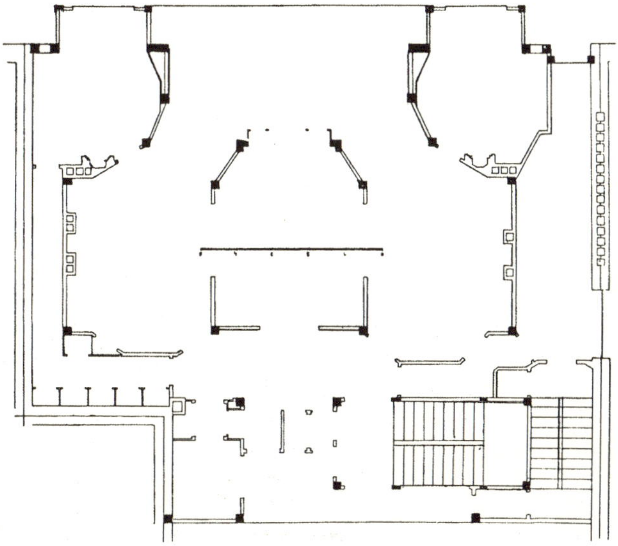

Fig. 70. A. G. PERRET: RUE FRANKLIN, 1903. PLAN of a Story
Note the few concrete piers on which the slender building rests. Staircase with glass-brick walls to the right.

Fig. 71. A. G. PERRET: The PERRET STUDIO on the Ground Floor of rue Franklin, 1903
The *fenêtre en longueur,* later used by Corbusier, is developed here already in 1903 as an interior plate-glass partition wall.

Fig. 72. A. G. PERRET: GARAGE RUE PONTHIEU, 1905
First logical attempt to let the FERROCONCRETE CONSTRUCTION become facade. The reminiscence of Viollet-le-Duc's rose-window Gothic in the splendid, vast central opening is completely irrelevant.

word: the opening! These were also taken up in recent French architecture and independently developed further. Perret himself judged his building: "The (world's) first attempt in the aesthetics of reinforced concrete." At the same time, this building of modest proportions begins a series of large garages that in today's building activity in Paris are almost the only places where positive design can be discerned. **Figs. 125, 139**

The new constructional interpenetration of a building is also found in Perret's Théâtre des Champs-Elysées of 1911–1913. The significance of this building is seen less in its traditionally attired space than in its innards, its dissolution into a concrete skeleton. In this very complex building—two theaters, one of which is suspended above the vestibule—the whole interrelated play of forces is transferred to the static calculations of load and support. Certainly the facade, like the auditorium, is designed in a conventional manner, and, just as in many American skyscrapers, the logical relation between skeleton and flesh is not achieved. But here as there the integrity of the frame still possesses unlimited possibilities for development. These possibilities are more valuable than any—for the moment—aesthetically satisfactory design. **Fig. 73**

The Casablanca docks (1916) and the church at Le Raincy (1922/23) are assembled with the same design elements: the Casablanca docks are of extreme lightness. The shallow vaults of the roofs are—for the first time, as is emphasized—executed with a membranelike thinness (3 centimeters). A few perforated layers of open bricks provide ventilation and animation. **Fig. 76**

The church of Le Raincy (1922/23) employs the same system:[75] flat vaults only a few centimeters thick form the three aisles. Expression and animation arise mainly from the construction. The nave has a longitudinal barrel vault, the side aisles spring with small transverse vaults against it. The lateral thrust is thereby effectively eliminated, and the room appears elastically modulated instead of rigid. **Fig. 75**

The exterior skin is only slipped on. Accordingly, it is perforated all over. The Casablanca ventilation system is—somewhat ornamentally, ceremoniously—utilized here. The form of these concrete stretchers in the windows may be questionable, but not their function: they serve as effective light filters. They block senseless streams of light and overdimensioned stained-glass windows that contradict the clarity of the concrete design.

The system of support consists of only four columnar rows.[76] It acquires an expressive quality because the outer rows are slightly set off from the wall.

The space breathes a lightness that is achieved only by an inner transcendence of matter.

The most recent generation carries on directly from Perret's achievement because—besides the academy—in him survives the legacy of the French constructors of the nineteenth century. He has the power to integrate a material—ferroconcrete—into the organism of architecture on a constructional basis.

[75]) While it should not be forgotten that Anatole de Baudot in 1894 erected the first concrete church—Saint Jean de Montmartre—Perret is to be credited with the first actual constructional reconsideration of the material. **Fig. 77**

[76]) The exterior of the church, especially the tower, is not free of a certain "concrete Gothicism," which is also manifested in the church of Sainte-Thérèse à Montmagny and above all in the design for the basilica of Jeanne d'Arc (Jamot, pls. 47, 48). But even in this design the boldness of the construction of a glass-and-concrete tower 200 meters high should not be forgotten. Perret worked with double-glass walls, as did Corbusier in his subsequent design for the Assembly Hall of the League of Nations building.

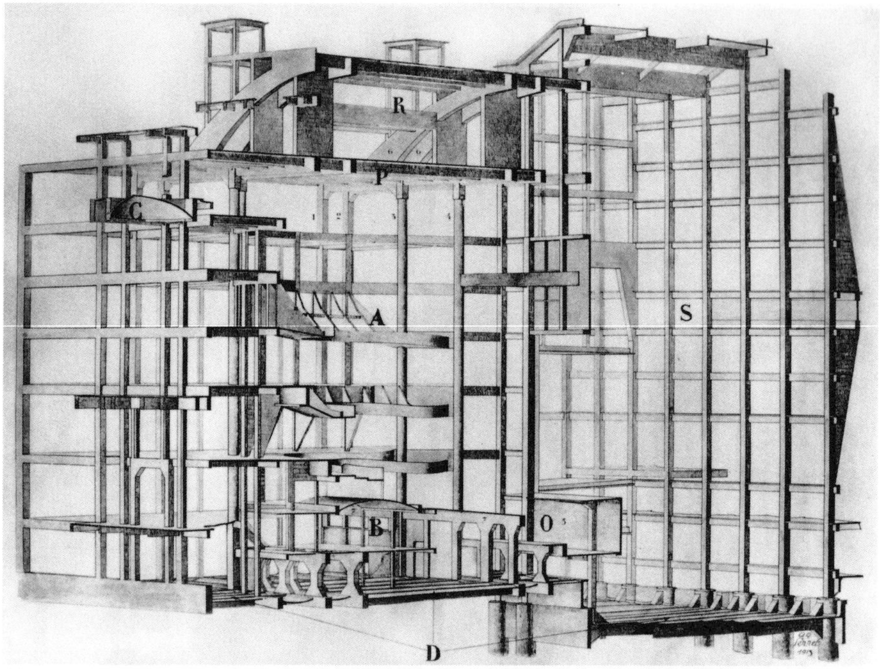

Fig. 73. A. G. PERRET: CHAMPS-ÉLYSÉES THEATER 1911–13
View into the skeleton frame. Note the continuous ferroconcrete columns on which the building is hung.

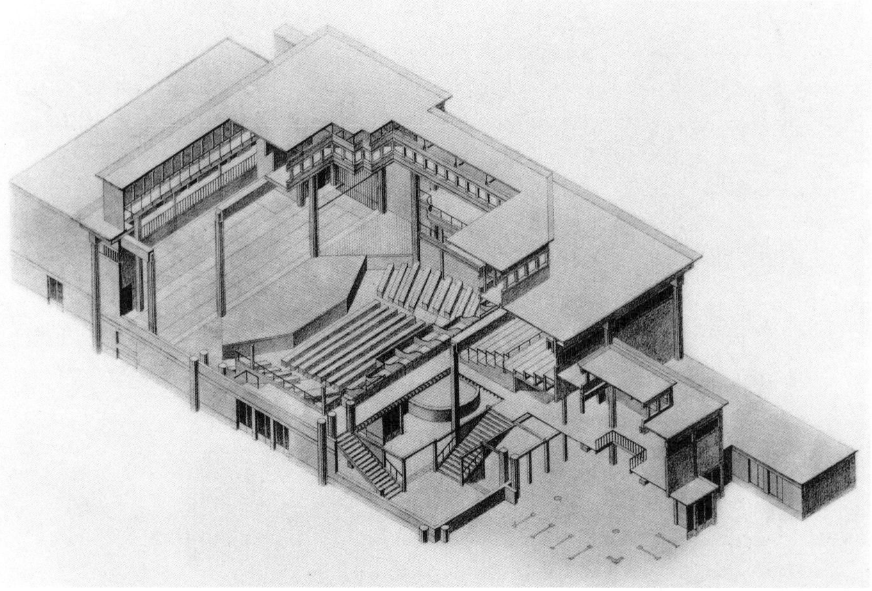

Fig. 74. A. G. PERRET: THEATER OF THE PARIS DECORATIVE ARTS EXHIBITION, 1925
Not the tripartite stage that has provoked the authorship conflict with van de Velde but the remarkable combination of wooden columns with ferroconcrete and iron appears to us to be what is worthy of further development in this building. Wood is a material that has found its new position in contemporary architecture from a very fragmented beginning.

Fig. 75. A. G. PERRET: CHURCH OF LE RAINCY 1922/23
The Casablanca docks and the church of Raincy are based upon the same formal elements.

Fig. 76. A. G. PERRET: Casablanca Docks 1916

Fig. 77. ANATOLE DE BAUDOT: SAINT JEAN DE MONTMARTRE, PARIS, Begun 1894
First church with ferroconcrete frame. At a time when only the most hackneyed patterns were used for representational buildings, it was an extraordinary risk to use a constructional FERROCONCRETE framework as an essential component of a church. By using ferroconcrete, DE BAUDOT's costs were approximately half those of other entrants in the competition (cf. p. 96). Cf. A. de Baudot, *L'Architecture et le ciment armé* (Paris, n.d.).

Perret, *architecte-constructeur,* is inspired by the material. He perhaps accomplished for ferroconcrete what Henri Labrouste accomplished for iron. In his language of forms Perret stands on a pathetical pedestal from which none of his European contemporaries can rid themselves.[77] Moreover, the classical canon survives in the works of the French architect, just as [Jean] Racine, Molière, and [René] Descartes live on in the French people.

Lloyd Wright is not inspired by the material like Perret. He is no pathbreaking constructor. But he lives on American soil and, despite all romanticism, is better able architecturally to design an unpathetical, self-evident way of living.[78]

[77]) It is to be thoroughly appreciated that the importance of the *architecte-constructeur* would diminish significantly should one classify it by form or by its use of ornament. It is significant that architectural expression becomes more consistent wherever a certain material spareness occurs and buildings are erected with the absolute minimum: utility buildings, 25 bis rue Franklin, the church at Le Raincy. The consistency of the design is easily obscured in those instances where unlimited means are available: the Champs-Elysées Theater, the basilica of Jeanne d'Arc, and the skyscraper projects of 1922. Likewise in his project for the League of Nations building.

[78]) A striking example of Perret's knowledge of how to design an open transition from house to garden is provided by a villa in Versailles of 1926 (fig. 110).

TONY GARNIER

Fig. 78. TONY GARNIER: Design for the QUARTIER DES ÉTATS-UNIS. LYONS 1920
The elements for the types of houses are already contained in a similar way in Garnier's *Cité industrielle* of 1904: no light wells, accessible roofs, open staircases, and in the house plans one large living room with cellular bedrooms.

Perret adopts the French tradition of penetrating a building constructionally, Tony Garnier (born 1869), above all, that of urban organization. The roots of his life's work, too, lie shortly after 1900.

He begins as a *pensionnaire de l'Académie française* in Rome with a study: *Cité industrielle.* That is not an acceptable project for a prizewinner of the academy, and to mollify his masters in Paris, Garnier chooses to reconstruct "Tusculum."

La Cité industrielle:

In the configuration of the site (terrace landscape, large river valley), the large-scale project for an industrial city of 35,000 residents resembled the architect's hometown: Lyons.

According to Garnier's statement, the whole project (*l'ensemble*) was exhibited in 1901, its details in 1904.[79] For all of its formal restraint, the project instinctively anticipates tasks that the new architecture would realize over the course of the next two decades. The picture that Garnier sketches is subsequently confirmed by life, for it is not slapdash visions that, like the highs of cocaine, dissipate already the next day, but it shows the engineer-minded penetration of the detail. The results were thus the product of a precise presentation of the problem and attention to the smallest cellular detail. Garnier proceeds: (1) from the building material, (2) from the urban organization. A social conviction leads to the idealistic superstructure.

From the start, the building materials are concrete and ferroconcrete, still

[79]) Published by Vincent (Paris, 1917).

Fig. 79. TONY GARNIER: Design for the Underground SUBURBAN RAILROAD STATION of the CITÉ INDUSTRIELLE 1904

In the age when—forsaken by all instincts—one hid a railroad station beneath marble architectural trappings, GARNIER as the lone forerunner designs his railroad station by taking into account function and the new materials: glass and concrete.

unusual in housing at that time. "The foundations and the walls consist of concrete; the ceilings, the roofs, and all important buildings almost exclusively of ferroconcrete." Even if in some cases the houses in their details cannot cast off their cubic unity: in accordance with the idea, the consequence of ferroconcrete construction is already drawn:

These houses have no cornices, the roofs are accessible, and one already works with differences in levels. For Garnier, the flat roofs are a completely self-evident result of the construction.[80] Freely suspended concrete stairs lead up to the roofs of multistory apartment houses—in part already covered like a railroad platform. The city—seen from above—receives its uniform flow from the horizontal fabric of roofs and the steplike ascent of the iron constructions of the factories and warehouses.

Garnier is already working with differences in levels. His schools have open and "covered" lawns (*préaux couverts et découverts*) that run freely beneath the terraces like the lawns beneath the chambers and the secretariat of Corbusier's design for the League of Nations. From these differences in levels emerge new tensions and interrelations of surfaces, solutions that Garnier himself was never allowed to translate into reality.

The house, which always proved to be least accessible to new formulations, here, too, remains restrained in its classicist form. The full consideration of each house, however, is based on principles that only today are taking effect: bedrooms oriented toward the south, "large and small courtyards, which means spaces that are enclosed by walls and are meant to provide light and air are precluded. Every room, however small, must be illuminated and ventilated from without."[81]

The simpler the design, "the lighter will be the construction and consequently the lower the cost. The simplicity of the means leads logically to a great simplicity of expression (*une grande simplicité d'expression*)."[82] Public buildings in the *Cité*

[80]) It should also not be forgotten that the classicist tradition is closely related to the flat roof, and formal connections lead to it.

[81]) Preface to the *Cité industrielle.*

[82]) Preface to the *Cité industrielle.*

industrielle display a greater informality. Already in the first half of the nineteenth century, hospital and prison complexes in France were broken up into pavilions in order to avoid light wells. Fig. 92

The assembly halls, the SUBURBAN RAILROAD STATION with underground tracks running through it and a renunciation of monumentality as later designed by Mart Stam and the young Swiss,[83] display the same informal character as the gardenlike treatment of city streets around schools and the connection of single houses whose plots flow into one another.[84] Fig. 79

Under Edouard Herriot, Garnier found the opportunity in Lyons to realize some of his projects. His large concrete complexes: stadium, slaughterhouse, hospital, define Lyons's townscape. What he had begun in his utopian *Cité industrielle,* he continues in his second published work, of 1919. The *Grands Travaux de la ville de Lyon* presents a curious mixture of works executed, to be built, or in progress.

Garnier's buildings dragged on due to the war. The slaughterhouse, begun already in 1909 and largely finished in 1913, is now almost completed. Figs. 54,80,81

The Grange-Blanche Hospital (project and drawings from 1911, begun in 1915), is probably not so far along. The stadium, from 1916, was realized most rapidly.[9] Figs. 82,83

The most difficult buildings to execute—as always in France—are residential buildings. Only five houses now stand in a forgotten suburb of a large district (Quartier des Etats-Unis, designed in 1920), through which should have flowed an avenue with open gardens 50 meters wide by 5 kilometers long. Even these houses suggest solutions:[85] the greatest possible suppression of vestibules, small kitchens, one large living room, small bedrooms.

Garnier is undoubtedly impeded by classicism, particularly in the stadium, but aside from these Hellenistic aspirations, he possesses what Edouard Herriot praises in his preface to the *Grands Travaux: "Une méthode rigoureuse."* The clearest indication of the crossing of classicist, monumental elements with the future-oriented attitude is found in the twenty-two pavilions of the Grange-Blanche Hospital. It is regrettable that in the final design Garnier did not relax the original, somewhat rigid axial arrangement of 1910. The individual pavilions are built of heavy concrete masonry with high-ceilinged infirmaries. Given the skill with which concrete is handled in France, Garnier can venture to form the amphitheatric lecture halls of the individual pavilions in concrete, including desks and bleachers.

Elevators carry the convalescent in his bed to the generous terraces to which the roofs have been converted. Despite the heaviness of particular details of the design, the eye, through the interplay of the various horizontal surfaces, has an impression of the air always separating and hovering, just as our future cities will be shaped.

A survey of Garnier's work indicates that the utopia of his *Cité industrielle* remains perhaps his most important contribution. A fantastic expansiveness grows out of a cellular interpenetration. One senses the combination of ratio and vision that will perhaps most clearly silhouette the coming age.

[83]) Designs for the Geneva-Cornavin railroad station by Stam and by Hans Wittwer (Basel), 1924. A project by [Julien] Flegenheimer that is not worthy of mention will be executed.

[84]) Mies van der Rohe attempted something similar with his first design for the Stuttgart Werkbundsiedlung of 1927, but it was not accepted.

[85]) The designs can be found already in the book *Grands Travaux,* therefore before 1919, and show how problems mature at the same time, in different places, independently of each other (cf. Holland: Oud's plan for Tusschendijken, Rotterdam).

Fig. 80. TONY GARNIER: LYONS SLAUGHTERHOUSE 1909
Largely completed in 1913. To the left, the large hall of the cattle market (cf. fig. 54). In the background, the actual slaughterhouses. (One of the first aerial photos in existence [1913], for which we can thank the generosity of the architect.)

Fig. 81. TONY GARNIER: LYONS SLAUGHTERHOUSE
Covered passageway in ferroconcrete, which has an elevated track—"conveyor belt"—on which the meat is transported. It is one of the four tongue-shaped connections clearly seen in the background of the aerial photo.

Fig. 82. TONY GARNIER: LYONS STADIUM 1916

Fig. 83. TONY GARNIER: LYONS STADIUM 1916

Fig. 84. TONY GARNIER: ROOF of one of the 22 PAVILIONS of the GRANGE-BLANCHE HOSPITAL, LYONS. Construction Begun 1915
Despite the massiveness of its form, the harmony of the many airy islands conveys a hint of what our future cities will look like.

Fig. 85. TONY GARNIER: GRANGE-BLANCHE 1915
The same pavilion with a view toward the mouth of the stairwell and elevator shaft. (During construction in January 1927.)

Yet Garnier's buildings still carry the mass and weight of a pathetical attitude toward life. The influence of the academy is not the only thing to blame. The conflicting nature of his work is a characteristic of his entire generation. As is typical of the legacy of the nineteenth century, superficially representational and future elements are often placed immediately beside each other.
In drawings of recent years, Garnier dreams nostalgically of distant times in southern laurel groves. Just as from another perspective Lloyd Wright lately designs fantastically shaped "residential yachts," or houses that burrow back into the folds of the earth. These are constantly recurring features of a generation of architects in whom the divided soul of the last century survives: historical burden and constructional penetration.

LE CORBUSIER AND THE YOUNGER GENERATION

Destiny of a generation: the past generation recognized and consciously formulated the importance of the place of work. The value of utility buildings (factory, warehouse, office building, the beginning of the problem of the tall building).
To the present generation falls the task of lifting the HOUSING PROBLEM out of individual dilettantism and pseudohandicrafts and onto the level of industrial standardization through the most precise and thorough consideration of housing functions. The future generation—which is already emerging from the mist—will once again prefer large construction projects. But this time variable buildings, open to all possibilities and closely connected with the means of transportation. With elevated and underground trains, airplanes, waterways. Comprehensive relational coordination of all means.
LE CORBUSIER, the French-Swiss from La Chaux-de-Fonds (born 1887),[86] gets credit for having placed the housing problem, the notably most underdeveloped aspect of French architecture, unequivocally in the center of his theoretical and practical activity. Corbusier, in any case, appears certainly to have loosened the tongues of many a young Frenchman. If we were today to attempt to describe his function, we would immediately be faced with the two-part question: where does he stand within the French movement, and by what means has Corbusier advanced the housing problem beyond the inherited norm? Only then can we discuss the formal structure of his buildings.
Corbusier depends entirely on the tradition of ferroconcrete. Only in France is it possible for an architect to trust this material unconditionally. The corrosive "ifs" and "buts" with which this material, often legitimately, is being opposed in other countries especially by leaders of the avant-garde, and which cause insecurity in production, do not apply in France. One need only think of Holland where, despite the best intentions, residential construction cannot be freed from the use of brick. The city of Amsterdam has allowed concrete villages to be erected according to the most varied methods without achieving any real satisfaction. Robert van 't Hoff's concrete-masonry house in Huis ter Heide (1915), which, by the way, is not a true "skeleton **Fig. 86**
house," remains a process of construction virtually without followers, although in a formal sense it became a point of departure for the whole Dutch movement. Corbusier

[86]) Further biographic data may be found in our article on Le Corbusier in volume 22 of the Thieme-Becker *Künstlerlexikon*.

has principally mastered two things: the ability to simplify things to an often almost dangerous terseness, and an unswerving consistency in development. Since his first house in the Swiss Jura (1916),[87] which is externally of conventional form, but already Fig. 87 contains the germ of later developments, Corbusier has composed his buildings with a ferroconcrete skeleton.

Perret's work contains prophetic germs and constructionally ingenious individual solutions: that of a precursor. No stylistic comparison, just this: when one visits Perret, he likes to hold up the flawless, self-contained oval of an ostrich egg, which he considers to be the most perfect earthly form. The only "nature" in Corbusier's Pavillon de l'Esprit Nouveau (1925) is that of an eternally open, self-transcending mother-of-pearl spiral of a large turbinate shell. Like no one before him, Corbusier had the ability to make resonate the ferroconcrete skeleton that had been presented by science. We do not mean his designs. We mean the skill with which he knows how to translate construction, the frame, into the new housing function. Out of the possibility of hanging the whole weight of a building on a few ferroconcrete pillars, of omitting the enclosing wall wherever one so desires, Corbusier created the eternally open house.

All of his architectural solutions lead back to it: his city on concrete piers (1915), the suspended houses that toward the base appear to become ever lighter, the cubes of air that spill over into his apartment houses (the first of these buildings, which are essentially villas set on top of one another, is to be built in Frankfurt), the gardens on the roofs and sides. Cubes of air within, cubes of air without. Cubes of air down to the very smallest units at Pessac and the individual cells of a *cité universitaire*. Maximum of air, minimum of walls!

This flow of air through the house: inside, outside, below, above (the flat roof is but

HOLLAND 1915

Fig. 86. ROB. VAN 'T HOFF: House in Huis ter Heide near Utrecht 1915. (Concrete)
Independent treatment of Lloyd Wright's influence. The horizontal concrete slabs are attached to the house like wings, without the cube itself being permeated by cubes of air.

FRANCE 1915

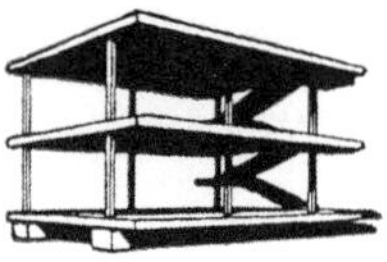

Fig. 87. LE CORBUSIER. Ferroconcrete Skeleton Frame for the Housing Settlement "Domino" 1915
C. develops the new housing function from the ferroconcrete skeleton frame with a thrusting boldness that has enriched all of architecture. From the elements of concrete pillars Corbusier develops the suspended, open house and bestows on it a previously unknown, exhilarating lightness.

87) Pictured in *Kommende Baukunst*, p. 62. We might also here call special attention to the German translation of Corbusier's best book, *Vers une architecture*, which the translator, Hans Hildebrandt, has published under the title *Kommende Baukunst*, more richly illustrated and more carefully printed than the original (Stuttgart: Deutsche Verlagsanstalt, 1926). [Translation of] Corbusier's *Urbanisme* is also in preparation.

a partial problem in a larger unity): this is what we demand from the new house! The reinterpretation of lean ferroconcrete construction in the redesign of the house demanded by the age and its will is what we call the true productive aspect of Corbusier's achievement.

Why should the house be suspended and made as light as possible? Only thus can we put an end to that fatal legacy of monumentality. As long as the play of load and support, in reality or symbolically intensified (Baroque), received its meaning through load-bearing walls, weight was justified. Today—with nonsupporting exterior walls—the ornamentally accentuated play of load and support is an embarrassing farce (American skyscrapers).

We will eagerly pursue the extent to which the house can be suspended without falling into lyricism. Corbusier has been reproached because his houses, especially those in Pessac, appear as thin as paper. In fact, they avoid supports projecting from the wall, cornices. The solid volume is opened up wherever possible by cubes of air, strip windows, immediate transition to the sky. The new architecture shatters the original conceptual polarity: space or plasticity. The new situation can no longer be understood with these old terms!

Corbusier's houses are neither spatial nor plastic: air flows through them! Air becomes a constituent factor! Neither space nor plastic form counts, only RELATION and INTERPENETRATION! There is only a single, indivisible space. The shells fall away between interior and exterior.

Yes, Corbusier's houses seem thin as paper. They remind us, if you will, of the fragile wall paintings of Pompeii. What they express in reality, however, coincides completely with the will expressed in all of abstract painting. We should not compare them to paper and to Pompeii but point to Cubist paintings, in which things are seen in a floating transparency, and to the Purist [Charles-Edouard] Jeanneret himself, who as architect has assumed the name Le Corbusier. In his *Peinture moderne* ([Amédée] Ozenfant and Jeanneret, *Peinture moderne,* Editions Crès & Co.) he likes to assure us that he has deliberately chosen only the most ordinary bottles and glasses, that is, the most uninteresting objects, for his pictures so as not to detract attention from the painting. But the historian does not see this choice as accidental. For him the significance of this choice lies in the preference for floating, transparent objects whose contours flow weightlessly into each other. He points from the pictures to the architecture. Not only in photos but also in reality do the edges of houses blur. There arises—as with certain lighting conditions in snowy landscapes—that dematerialization of solid demarcation that distinguishes neither rise nor fall and that gradually produces the feeling of walking in clouds.

Fig. 88. JEANNERET 1924
Just as transparent objects interpenetrate in the painting, so Corbusier with every means also lightens the traditional gravity of the house.

Individual houses do not satisfactorily demonstrate this: housing settlements. The relation of several houses to each other. Corbusier has his first housing settlement behind him: PESSAC near Bordeaux. (Completed 1925).[88]

ORGANIZATION: Pessac is the first housing settlement in which no restraints were placed on the architect. So far, some fifty houses have been erected, but fifty more are planned.

In these relatively few buildings the advantages of standardization already became obvious. Room dimensions were normalized. Module: 5 meters. Rooms: 5 × 5 meters, or the subdivision 5 × 2.5 meters. Each building, according to its size, contains six, eight, or more cells. Window sizes derive from individual room sizes: 5, 2.5, 1.25 meters: ¼, ½, ¹⁄₁ windows. This standardization permits the greatest variety of applications while at the same time requiring but one order from the factory. Roof and floors have the same module: 5-meter concrete beams. In addition, there are also uniform doors, hearths, fireplaces, stairs.

Machines fabricate the insulating cinder blocks and the concrete beams on-site. I have personally seen the Ingersoll-Rand Company's "cement gun" (the concrete cannon) spray forth a 100-meter wall in just a few days. This aroused the resistance of the workers to the point of sabotage.

Fig. 101 For the time being, six different house types are being used. Of these, the *gratte-ciel* type (duplex) and the smallest type are repeated most frequently and thus determine
Figs. 94 to 97 [93 to 96] the overall look of the housing settlement. "Type 14" appears to us to be the most elegant: it most decidedly leads to that lightening of the dwelling. It is suspended on its concrete columns. The staircase is organically pushed to the outside in order to leave all space free for living. From that it follows naturally that staircases resemble railroad overpasses.

88) The name of the patron Henri Frugès of Bordeaux will not be forgotten in the history of the new architecture. Frugès told the architect, whom he had looked up after reading *Vers une architecture:* "You have my permission to put your theories into practice and to take them to their ultimate consequences. Pessac must be a laboratory. You have my full permission to break with all conventions and to abandon traditional methods" (cf.: *Science et l'industrie* 154 [Paris, 1926]). Cf. also Corbusier's statements in *Architecture vivante* (published by Morancé, 1927). Further on Pessac: [Steen Eiler] Rasmussen, in *Wasmuths Monatshefte,* 1926, no. 9.

Fig. 89. LE CORBUSIER: PESSAC. The Smallest House Type (Plan: 49–54)
In the foreground, small animal stalls are skillfully disposed along the wall. *Gratte-ciel* in the background.

Fig. 90. LE CORBUSIER: PESSAC. Smallest House Type with Barrel-Shaped Store Rooms (Plan 61–64)

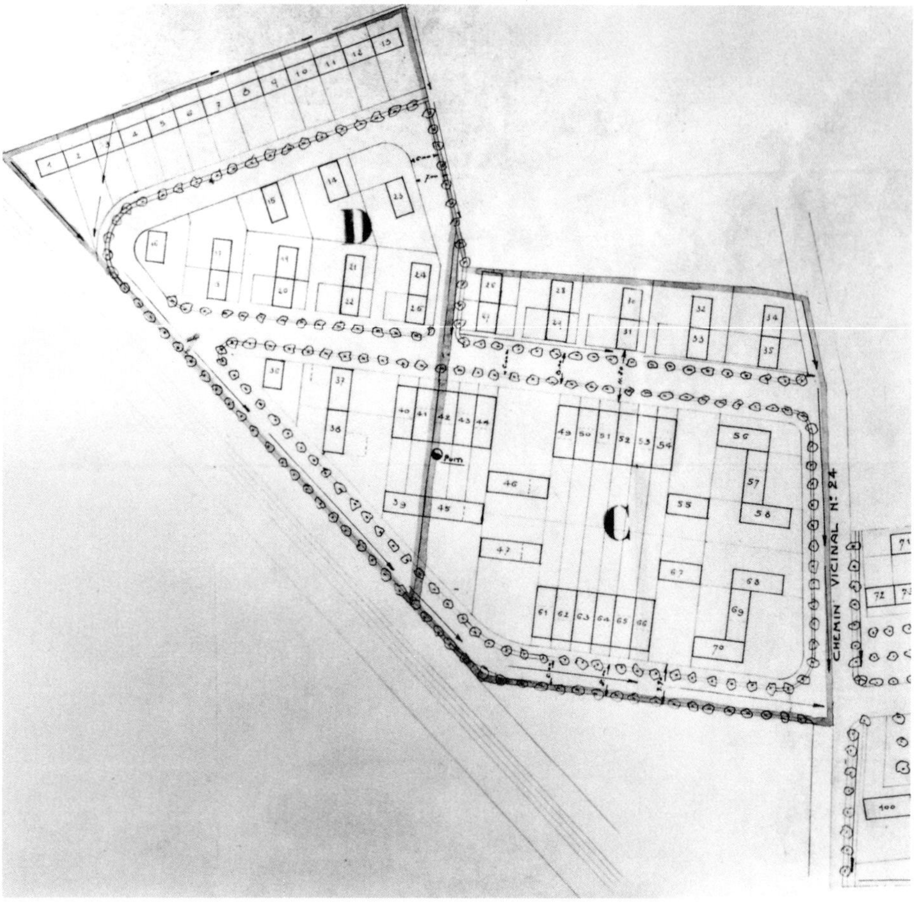

Fig. 91. The Executed Portion of CORBUSIER'S HOUSING SETTLEMENT IN PESSAC 1925/26

Liberation from the inflexible constraints of the housing settlement plans. The free, informal distribution corresponds to the character of the contemporary house, which is shaped so that proximity does not also mean crowdedness, for individual parts and the whole are opened up.

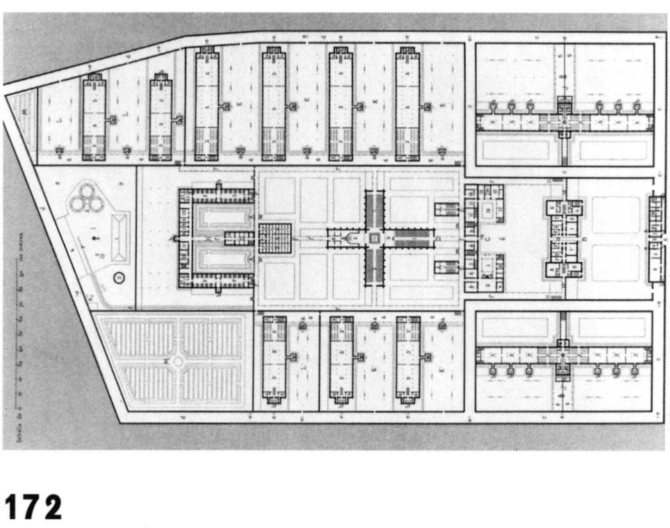

Fig. 92. [E.] LAVEZZARI: Prison. 1877

The informal floor plan also has a long tradition—especially in France. It is common in hospitals, schools, and institutions after midcentury.

Fig. 93. LE CORBUSIER
Pessac: "Type 14." Roof Garden.

Fig. 94. "Type 14." Stairway
In order to gain space in this southern climate, the stairs are placed outside the house. A detail of the anonymous design of the engineer: A RAIL OVERPASS becomes an architectural member.

Pessac: "Type 14." View of first story.

Figs. 95 and 96.

Pessac: "Type 14." Plan of first story.

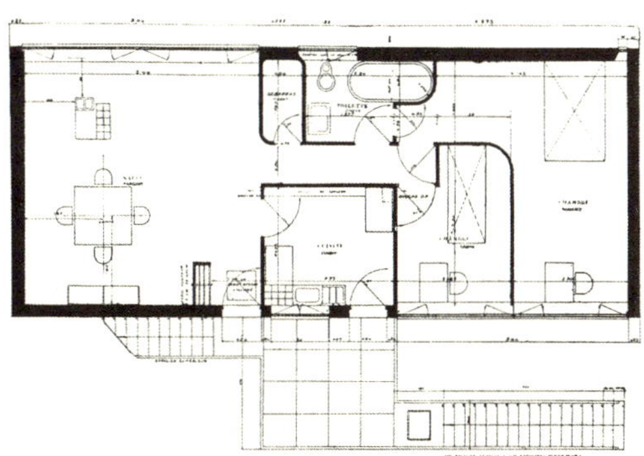

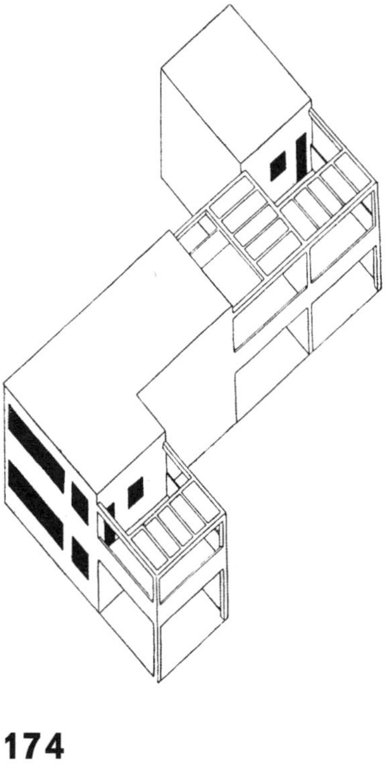

Figs. 97, 98, 99. LE CORBUSIER: PESSAC
Figs. 97 and 99. Three-Family House (Plan 68–78)
In the Middle: Three-Family House (Plan 56–58)

These Z-shaped houses are placed at the head of the current development (color: light blue).
Combining standard elements guarantees a quite unusual flexibility of the houses and a new profusion of potential views.

Fig. 100. LE CORBUSIER: PESSAC. The *Gratte-ciel* (Plan 17–35)
These taller structures are arranged so as not to interfere with the others. In order to preserve the unity of the whole, one is no longer bound by the facade plane or the frontal height.

Fig. 101. LE CORBUSIER: PESSAC 1925/26. (Type: Plan 1–13)
The spaces between the houses become roofed gardens. The vaulting is achieved by the least expensive means. Wire netting is sprayed with a concrete gun.

Fig. 102. LE CORBUSIER: STUTTGART 1927

It must still be noted that the basic elements of the Pessac housing settlement—including its 5-meter module—date back to 1921. One will find them in: *Vers une architecture.*

The flat contours of Pessac merge with the sky: the suspended canopies over the roof gardens form the transition. The color scheme is taken from Jeanneret's paintings:
Fig. 88 ethereal sky blue and light green, a more intense brown.

The interplay of the units can be judged neither spatially nor plastically. Only relations count. Relation of mass to void. Relation of smooth surfaces to perforated ones, relation of horizontal layers to vertical bodies. COLORS serve to lighten the volumes, to advance and recess surfaces.

How could one judge the space and plasticity of, for instance, the brown row houses of the smallest type without taking into account the oscillating relations between things? These houses that so rigorously respect the planar surface are themselves being penetrated with expansive, onrushing cubes of air, which among themselves receive new stimulation and modulation—as by the swelling, visually hard-to-discern vaults (pantries). The row houses as a whole again reach into the space next to and behind them. Still photography does not capture them clearly. One would have to accompany the eye as it moves: only film can make the new architecture
Fig. 90 intelligible! But even then only in a limited excerpt: does one really think that the wall on the right, as taut as a movie screen and altogether deprived of its corporeality, stands there only accidentally, unrelated to the opening and surface of the brown elements next to it? Certainly this effect in its particular features is not calculated. But the result evolves by itself from the elements of an architecture that—freed from the play of load and support—has cast off the anthropomorphic shackles. We owe it to the Dutch, to [Piet] Mondrian and [Theo van] Doesburg, that they first opened our eyes to the oscillating relations that may arise from surfaces, lines, air.[89]

[89]) In his theory Corbusier is often less daring than in his design. In *Vers une architecture* there is a chapter on *tracés régulateurs*—Hildebrandt translated it [into German] as *Aufrißregler* [regulators of the elevation], that is, the attempt to overlay similar triangles on the facade. Berlage was perhaps the last—1907—to be allowed to do that. This is permissible for a uniformly elaborated, anthropomorphic architecture. But in the case of Corbusier it is absurd. The proportions that will result from standardization are no longer restricted by the old formulae, but apart from this, these proportional formulae are invalid today because a BUILDING is no longer a closed form like a Renaissance palace but demands CONNECTION TO THINGS NEXT TO IT. Perhaps Corbusier himself has negated his method in the LA ROCHE house; there the proportional formula is inscribed in the facade, whereas the functionally projecting wing for the art collection naturally refuses to be restricted by a schema of similar triangulation.

Figs. 103, 104. LE CORBUSIER: MIESTCHANINOFF HOUSE. BOULOGNE-sur-SEINE 1923/24

A step on the way to freeing the residence from its innate weight. The flat roof is still not organically integrated into the body of the house, and the bridge-shaped veranda passage is too literal, but the rigid volume of the house appears already under attack.

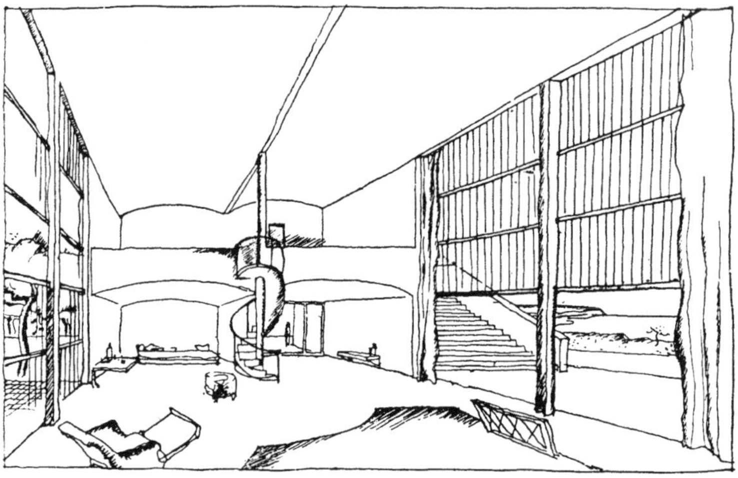

Fig. 105. LE CORBUSIER: Villa on the Sea 1921
Without having found an actual architectural form, these designs already contain the vertical fusing of space, broad openings, and the greatest possible avoidance of partition walls made possible by the FERRO-CONCRETE SKELETON.

Figs. 106, 107. LE CORBUSIER: La Roche House in Auteuil. Architectural Elaboration of the Sketch in Fig. 107
Left: The structure.
Right: The finished building.

Fig. 108. MART STAM: Large Room in the Werkbundsiedlung House 1927
With the avoidance of window lintels with their shadows the last barrier stressing the division of interior and exterior falls away.

Fig. 109. LE CORBUSIER: COOK HOUSE 1926/27. Boulogne-sur-Seine
View from the study (cf. plan) to large room, stairs, and roof terrace.
Exterior space (roof terrace) and the various interpenetrating levels of the interior space are blended together. Fig. 105 and fig. 109 indicate the distance traveled in five years.

Fig. 110. PERRET A. and G.: Study in a Villa in Versailles 1926
Perret also throws the house wide open toward the garden. But integration remains limited to one level.

One realizes: normalization and standardization are only expedients for the liberation of architectural vision! They play no other role in architecture than the vacuum cleaner or the washing machine in the average household: to free the mind for better things!
Details lose their disastrous tyranny only when they are standardized. In every sense it can be understood for the future: there are no more details, there is only an ensemble!

What is it like INSIDE these buildings?
From the beginning, the will requires a strong interpenetration and interrelationship of all parts and connection to the outside:

Fig. 105 In 1921 the basic ideas were already formulated in the sketches for the book *Vers une architecture:* avoidance as far as possible of partitions. The spiral staircase as an open and interconnecting member.
In the well-known studio Corbusier designed for Ozenfant in 1923, the idea acquired form. Garage within the body of the house. A spiral staircase winds upward, connects to a hall and, in a further rotation, opens unexpectedly into the dining room. A suggestive notch allows one to make out further spaces. After a turn, one stands in the most beautiful studio, in which the work cell is suspended like a crow's nest. A place of isolation. All other rooms breathe together.
The La Roche house in Auteuil of 1924.[90] The spheres of the stories interpene-
Fig. 106 trate one another. Just as Lloyd Wright—twenty years earlier—fused the rooms of the house horizontally, so Corbusier primarily does it vertically.[91] It is good to peel the skin off the house and to expose in the naked skeleton itself how the structural parts flow seamlessly into one another. Spaces and light, interior and exterior stream together. In an unexpected spot in the ceiling (above the wall-less library of the sec-
Fig. 107 ond story) is cut a square skylight, through whose clear glass one sees the sky. Even the ceilings are light. No burdensome closure. (Rietveld, for example, was able to give his house in Utrecht skylights at a stroke.)

1926/27. In the La Roche house Corbusier only dared to put a lateral and parabolically advancing part of the house on posts so that the garden could continue underneath. For the Cook house in Boulogne-sur-Seine he draws the conclusion of setting
Figs. 11, 109 a house on posts and lets only a small portion (staircase-vestibule) be walled in. Only on the upper story do the freely supported walls begin, and there the continuous strip windows are displayed with full consistency. The roof garden of the La Roche house is, to be sure, already thoroughly modeled, but only here is the integration of house and roof complete. The large room of the house extends above the roof plane up to the height of the covering baldachin, which protects against rain. A staircase leads
Figs. 111, 112 from this large room to a smaller study which, at roof level, forms the middle link of an indivisible interlocking of interior and exterior.
Figs. 113, 114 1927/28. De Monzie house (Garches, near Saint-Cloud). Just as Corbusier in the Cook house fully disengaged the building from its socle, so in the de Monzie house he attacked the facade of the building with enormous boldness and left it so penetrated with air that one can almost speak of a crushing of the actual house volume.
Corbusier's utter mastery in accentuating certain lines in a fascinating way is also evident in his ability to conjure up spaces whose appearance far exceeds their real cubic dimensions. There is in these terraces and bridges an almost balanced manipulation of spaces and particles of space that continuously interpenetrate. There seems to be a limit here. One senses that the architect has not fully escaped the

90) Cf. our remarks, "Das neue Haus," *Kunstblatt* (April 1926).

91) The elements of the open, two-story plan are found in many studios in France. Corbusier has elaborated what was rudimentarily available. It remains an open question how such vertical interpenetrations may be applied to non-Latin peoples.

Fig. 111. LE CORBUSIER: COOK HOUSE. BOULOGNE-sur-SEINE 1926/27. ROOF TERRACE
The ROOF is completely integrated into the organism of the house. The house is no longer a structure whose gable happens to have been shaved off. It is opened on all sides and has finally lost the closure of cubic armor.

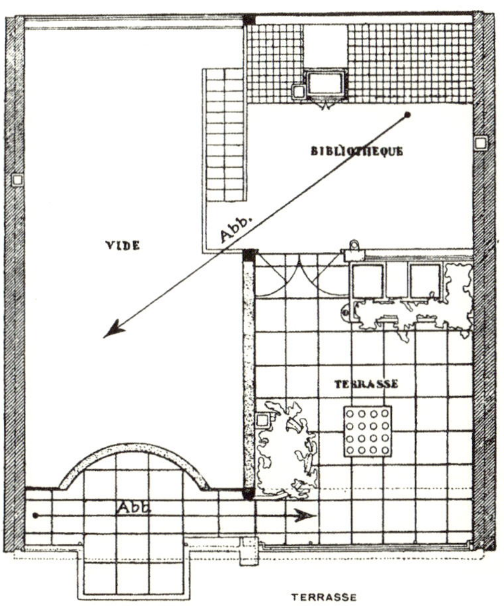

Fig. 112. COOK HOUSE. PLAN of Top Story
Interlocking of inner and outer, of roof space and interior space. Cf. figs. 109, 111.

Figs. 113, 114. LE CORBUSIER: DE MONZIE HOUSE in GARCHES near Saint-Cloud, 1927/28

Attempt to lighten the house not only at the base and upward but also at the front in order to let it be penetrated with air. Breaking up the tangible facade! The "Type 14" house at Pessac is the precursor of this deeply cantilevered terrace. Fig. 94.

Fig. 115. LE CORBUSIER: ASYLUM FOR THE HOMELESS, PARIS 1927
The architect has removed the ground floor to gain space.

danger that today is always posed by unlimited financial means.

One wonders if it is possible for those of us who are accustomed to a quick pace to cross such a terrace—so flooded with air as if by floodlights—without having the feeling: I am on a stage. At such points the next generation of architects begins to smell danger. We have no doubt that we must accept slow and gradual development, continually controlled by social and financial obligations, if architecture and our culture are to receive that perhaps still-unrealized unity and constancy that have already been proclaimed in many other fields.

Fig. 116. LE CORBUSIER: ASYLUM FOR THE HOMELESS, PARIS 1927
Today in the important places, as in all ages that felt themselves to be laying the foundation of a new culture, there is a noticeable UNITY of design:
Thus the FUNCTION a window has to fulfill must be so predominant that its form should remain the same whether in an asylum for the homeless, in a luxury residence (de Monzie), or in the League of Nations building.

Nevertheless it is important that isolated experiments, such as the de Monzie house, are made, for they help to loosen up the still-too-rigid means of expression in architecture. If it is at all possible to design luxury today—the actual problems lie elsewhere—it would be only the kind of luxury shown in the de Monzie house: luxury of air volumes whose interpenetration and harmony let the new way of SEEING become form.

We know that human housing must be shaped more sensitively than garages. But a perhaps justifiable fear of aesthetic emphasis persists, driven by the experience of a century. No one thinks in terms of a schematic machine-for-living (*machine à habiter*). But more important for us than aesthetics and poetry is it that the architect concern himself with the biological function of the house and thus help to combat the coarseness with which these things in many cases are still being treated today.

We do not doubt that at the very moment when these things furnish a solution, an equally obvious beauty will emerge from houses as from ships and airplanes.

This beauty becomes still more intensive, for it appears connected to our human functions.

Then this point will also be reached without emphasizing an isolated aesthetics, which Corbusier illustrated as the difference between the Parthenon and an automobile: inner movement!

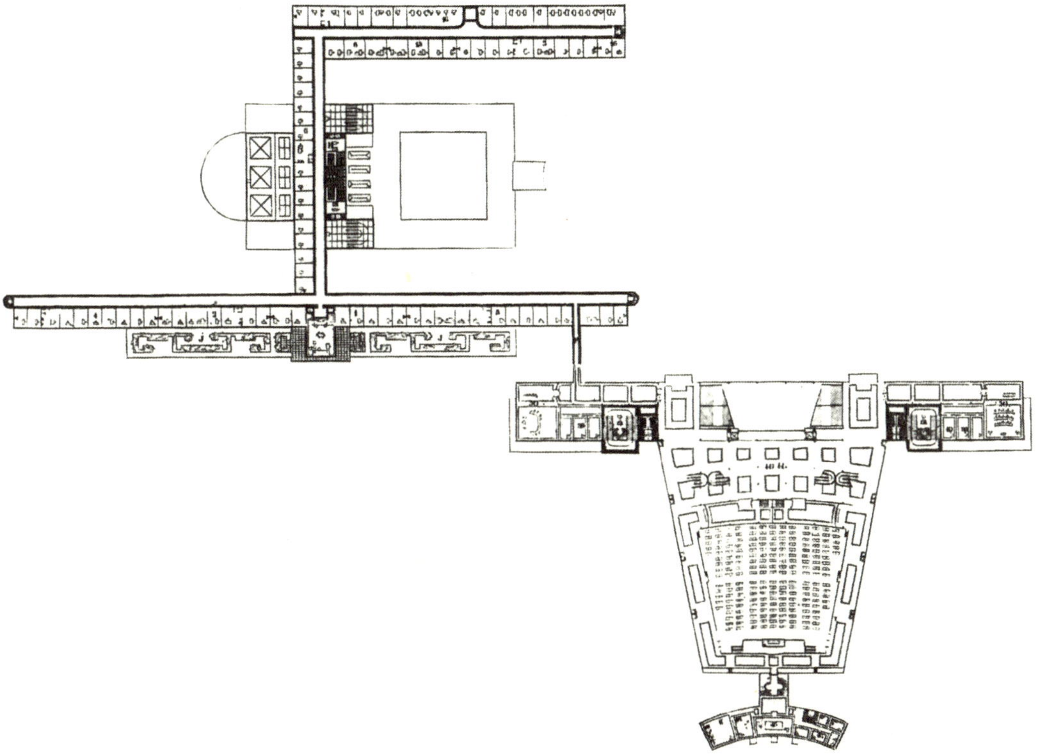

Fig. 117. LE CORBUSIER: LEAGUE OF NATIONS BUILDING 1927
ASSEMBLY BUILDING to the right, with the Presidential Pavilion in front. To the left, behind the assembly building and linked by passageways is the CENTRAL SECRETARIAT with library and archive projecting to the rear.

Fig. 118.
LEAGUE OF NATIONS BUILDING
Axonometric sketch. View from land toward lake and mountains. One notes the clear arrangement of the building complex without light wells. Covered roof terraces with gardens and broad vistas.
The secretariat is suspended on supports, beneath which flows the lawn. One sees the PLATFORM for arriving cars in front of the assembly building.

Fig. 119. LE CORBUSIER: LEAGUE OF NATIONS BUILDING 1927
The 140-meter-long PLATFORM (*quai de débarquement*) for arriving cars. This "railroad platform" is applied as a formal architectural device.

What is Corbusier's achievement? He has grasped the house with the sensitivity of a seismograph and released it from traditional weight.

He is not as precise a constructor as Auguste Perret, nor do his houses display, down to the last detail, the deliberate functionality of a J[ohannes] J[acobus] P[ieter] Oud. But Corbusier has ventured pioneering work like no one else in our time. He attempts to translate into the housing form that suspended equilibrium, that lightness and openness that iron constructions of the nineteenth century express abstractly. He has shown us how one must mold the house throughout—below, above, on the sides—in order to relieve its weight.

Corbusier's means are not unique to him but originate earlier in industry. The posts that support his houses can be seen in many French warehouses; the strip windows—*fenêtres en longueur*—are commonly found in factories and wooden barracks as the result of construction, and the concrete column running freely through the stories is of the same origin.

Garnier has already realized the flat roof and Lloyd Wright the free plan, but Corbusier is the first to conceive a house so thoroughly that the weight has actually been taken away from it, and it has not remained a block whose roof has simply been shaved off. Not the least of his roles seems to be the fact that he, in overriding all inhibitions, has helped people go their own independent way.

Corbusier's houses can be attacked on several points. He has been reproached for his romanticism, as when he—for instance, when he started out—takes over a whole series of formal motifs from shipbuilding (figs. 103, 104). There is also a certain danger of a strong, aesthetic emphasis, which today's architects understandably fear. Architecture today—like nineteenth-century construction—proceeds gradually, for it must lay down foundations for a long time to come.

Others may appear who will purify Corbusier's work. What is laid down nevertheless stays, for it is controlled by an architectural vision, and its roots are torn from the flesh of the age!

The THEORETICIAN: Corbusier has exercised as much influence on our time as a theoretician (his impact in Latin America is similar to here) as he has by his buildings. For an age in which even the artist is allowed to make use of ratio, it is understandable that one can be at once both a popularizer and a creative individual,

just as in politics—for instance, Lenin—one can be a journalist, theoretician, and state leader all at the same time.
In the pages of the journal ESPRIT NOUVEAU, which Jeanneret coedited with Ozenfant (1921/24), one can find almost everyone who in every field was active in formation of the epoch.
Corbusier's effectiveness is explained by his ability to simplify things to a point of lapidary succinctness. His presentations are not made unpalatable by any half-thought-out philosophical claptrap. He reduces the problems to the clarity of film titles. Of course, the French language helps give unequivocal wording. Subtly differentiated complexes are often dealt with summarily, but people retain the basic ideas. The three introductory chapters of *Vers une architecture* on "eyes that see not... automobiles, airplanes, steamships" cannot be expunged from our age, and at the same time they form a bridge to the forgotten predecessors in the nineteenth century.

Corbusier's design for the LEAGUE OF NATIONS BUILDING (first prize *ex aequo*) contains the elements that constitute Corbusier's work and whose gradual development we here tried to suggest.
The synthesis of constructional and standardized elements,[92] which are welded together through artistic vision into a totality, might explain the privileged position it has achieved among the 377 submitted entries.
We need to consider here the competition for the League of Nations only insofar as it touches upon the large issues that run through the entire century.
We thus encounter, perhaps for the last time in history, the name of the ACADÉMIE DES BEAUX-ARTS.
The academy was—as we mentioned—renewed by Napoleon at the moment (1806) when he was preoccupied with giving a feudal veneer to the new empire. The tradition the academy maintains is in no way a generally national one: it is the mentality of feudalism! Because feudalism as a social system is dead, the academy, too, is dead. That is the reason why for the last hundred years the academy's name has appeared only when it has been a question of impeding development.
Nothing has changed since Labrouste, already in 1840, proved that the academy's training produced students with an incapacity to deal with life (see page 98).
Already fifty years ago the secretary of the academy, the vicomte Henri Delaborde,[93] publicly declared that the doors to the academy opened only "when genius was gradually in accord with everyone," that is to say, the academy inconsiderately reserves for itself the traditional right to sabotage young talents, to suppress their development as much as possible, only to assimilate them in their old age.
When the Eiffel Tower was being erected, the academy was its most zealous opponent, only it could do nothing against the government then in power. For thirty years

[92]) ELEMENTS: ferroconcrete skeleton with uninterrupted supports, non-load-bearing walls with continuous strip windows, absence of the plinth, and inclusion of attic space within the building complex.

[93]) The state of affairs may be expressed with almost cynical candor: "Sooner or later they find that they have no adversaries left and, without yielding anything of substance, they are more or less in agreement with everybody. This is how it has been in our century with painters such as [Jean-Auguste-Dominique] Ingres and [Eugène] Delacroix, with sculptors such as David..., with such a musician as [Hector]

the academy sabotaged Henri Labrouste, up to the final works of his students, and it later even suppressed a [Eugène-Emmanuel] Viollet-le-Duc; and just as it yesterday ostracized Auguste Perret, so it today ostracizes Le Corbusier.
No one outside France has yet concerned himself with the academy. But through the inertia of events or the mentality of political leaders it is a reality. This is why we must concern ourselves with it, however little it can be doubted that this feudalistic holdover will soon be swept away.

The task of designing a building in which all nations would assemble to maintain the equilibrium of the world is completely new and belongs to the time. To want to solve it any way other than with TODAY'S MEANS is dilettantism, for, as we will show, one is already compelled by the task subsequently to patch the new building technologies (contained in the projects of architects working out the problems directly) onto the most ridiculous stage-set buildings.
The League of Nations building consists of two parts: the administrative apparatus (General Secretariat). This is a continuously functioning office concern, an office building that will have to use all knowledge currently available to achieve a smooth functioning of the business within. The second part: the chambers for the world parliament. Because of its vast dimensions the Assembly Hall is by definition a task that cannot be resolved with "monumental" means. If one did not take the acoustic problem into account, the requisite dimensions would make it impossible to hear a single word.
In addition, there are integral elements of the overall composition: the regulation of automobile traffic—five hundred cars must be able to arrive within a few minutes—and the accommodation to the site.[94]
Fig. 118 [117, 118] Corbusier has oriented his building complex slightly toward the south, toward Mont Blanc. The Assembly Hall advances, the General Secretariat recedes. The entire complex blends easily into the landscape, it is integrated with the surroundings, and through its architectural treatment—the General Secretariat on supports, extensive use of glass—it avoids any awkward heaviness. As in all truly new projects,[95] automobile traffic is not a concern that disturbs the architecture but instead becomes a stimulus for design.
The means supplied by the age are evident everywhere, often unseen and unused, as

Berlioz.... And so, in the domain of architecture, the same has happened to M. Labrouste, whose name... once seemed to personify the spirit of extreme independence..." vicomte Henri Delaborde, secrétaire perpétuel de l'Académie, "Notice sur la vie et les ouvrages de M. Henri Labrouste, lue dans la séance publique annuelle du 19 octobre 1878" [p. 36].

94) A detailed account of the competition and the project itself is not possible here. We refer to our remarks elsewhere: "Die Architektur am Wendepunkt" (cf. *Neue Zürcher Zeitung*, nos. 1247, 1249, of 24 July 1927, as well as "Wer baut das Völkerbundsgebäude?" *Bauwelt* 44 [1927]: 1093). See also *Schweizerische Bauzeitung* 90, pp. 13ff. (text by Corbusier); p. 59 ("Über das Problem der Akustik," by F. M. Osswald); on the legal situation, p. 180; on the breach of contract by the League of Nations, p. 197. The course of the competition is presented best in Christian Zervos's essay "Qui bâtira le Palais des Nations?" *Cahiers d'Arts*. See also the fall issue of *Architecture vivante*, 1927, by Jean Badovici and Corbusier. Finally the "REQUÊTE" to the Société des Nations signed by MM. Le Corbusier and P[ierre] Jeanneret (Paris, 1928). A pamphlet protesting the unlawful proceedings of the League of Nations.

95) The uncrowned project by Richard Neutra presents the most thorough integration of the traffic problem.

it needs a visionary outlook to recognize them. To build a living architecture means to have at one's disposal the ability to recognize these means and to transform them into architectural design.
Hence, acoustics provides the foundation of the design of the large Assembly Hall. The architect translates into form the knowledge discovered by the specialist (in this case G[ustave] Lyon, Paris). The result: even new formal possibilities arise thereby. Competition for the new League of Nations building means: conflict between a living and a rigid, feudal architecture.

RIGID-FEUDAL PROJECTS	LIVING PROJECT
SITUATION: No consideration of the terrain. The park is rolled flat and serves as a socle for the architecture. Rigid axial alignment according to an incidentally preexisting building ([Henri-Paul] Nénot's project).	SITUATION: Organic integration with the surroundings with respect to function and landscape. Southern orientation.
SECRETARIAT: Is perceived as a cumbersome appendage, for it spoils the "monumental" effect. One of the selected entries ([Carlo] Broggi) arranges the secretariat like armor around the assembly building with some twenty light wells. No "monumental project" can do without light wells.	SECRETARIAT: The design grows out of the task: office activities. No light well. Even the lowliest clerk has a view of lake and mountains.
ASSEMBLY BUILDING: Automobile approach is neglected. Instead fountains and colonnades. Problems of acoustics: nonexistent.	ASSEMBLY BUILDING: Automobile traffic generates the "facade" (platform). Acoustics lead to new design.
COSTS: Two, three, four times the budgeted amount of thirteen million francs.	COSTS: Twelve and a half million francs. (Le Corbusier).

SUMMARY: When they were submitted, the "MONUMENTAL PROJECTS" utterly failed on every point: for the monumental gesture completely drowned out all functions. This is thoroughly understandable. The armoring with colonnades and column orders binds much too tightly, like a corset, and is too coarse to be able even remotely to fulfill the demands put forth by life today.
Even the diplomatic committee that selected the architects of the four worst designs has understood that.[96] The committee demanded fundamental changes that were taken almost point for point from Corbusier's project.

[96]) The intrigues, questions of prestige, connections that obviously decided the issue appear virtually impenetrable. Looking over the whole complex, one can understand why the eighty-year-old Frantz Jourdain, the architect of the Samaritaine (cf. p. 119), is said to have commented: "*C'est une seconde affaire Dreyfus.*"

This competition recalls Les Halles in Paris, where Baltard likewise laboriously had to patch his building together from the ideas of others. As we already mentioned, the Paris city councillors of 1850 saw—according to the judgment of their contemporaries (cf. p. 114)—monumentality only in the accumulation of stone masses. Unfortunately, in the eighty years since then, those in power in the League of Nations have not advanced one step further in their judgment.

Who works alongside Perret, Garnier, and Corbusier?
First, the in-between generation—age-wise between Perret and Corbusier—that did not fully develop. Two names: Rob[ert] Mallet-Stevens, Henri Sauvage. One must first say about both: they, too, have the tradition of ferroconcrete in their blood. But they did not achieve consistent formulations.
Rob. MALLET-STEVENS (born 1886) is the movement's *élégant.* He remains a formalist who drapes the new means over the old frame. His youthful impressions revolve around the Stoclet house of 1905 in Brussels (by Jos[ef] Hoffmann), which, by the way, represents the best realization of the Wiener Werkstätte Movement. Mallet-Stevens has a special relationship with this house. In his youthful work *Une cité moderne* (Paris: chez Massin, n.d.—around 1914—introduction by Frantz Jourdain)[10] that house and the dreams of the Wiener Werkstätte keep recurring. No overall vision as in Tony Garnier's *Cité industrielle* already around 1900. Details. Arts and crafts. Yet the French tradition of ferroconcrete lives on in him. The hall of his
Fig. 120 transportation pavilion at the Paris Exhibition of 1925 is actually suspended on only two supports, and all five stories of his Alfa-Romeo Garage (1925) are hung from two parabolic girders so that no supports obstruct interior traffic. The facade and shop of this frequently reproduced building exhibit a mixture of Dutch and Viennese crafts influence (cf. by contrast J. J. P. Oud's Café de Unie in Rotterdam).
For the well-to-do in Auteuil the architect built an entire cul-de-sac named after him: RUE MALLET-STEVENS. The most interesting aspect of these four-story, single-family houses (ferroconcrete) with a half-dozen bathrooms is the imperative with which our age, even if only faintly heard, strikes down all ostentation. There is still—
Fig. 122 in the largest house—that multitude of rooms that formerly seemed indispensable to wealthy households. But the stridency of capitalism is nonetheless slipping away. The ground floor is devalued. Where formerly the suite of rooms unfolded, now reside the chauffeur, garage, and domestic facilities. Everything has been pushed upward. On the first story: only remnants of the old ostentation but—compared to the total cost of the house—negligibly little. No millionaire would previously have tolerated that. It is clear that the house has no garret. In the basement there are special rooms for furs, paintings, baggage, and supplies.
A dumbwaiter runs all the way to the roof; to the most comfortably arranged roof that we have ever seen. A flat concrete baldachin cleverly suspended from broad chimneys protects the table linen from every last particle of soot! Then an adjoining tiled pantry. Rooftop pantry. Le Corbusier's astringency reinterpreted for the *gourmand-élégant.*

If evidence were needed that contemporary architecture and a "rich" style are incompatible: the houses on the rue Mallet-Stevens would supply it. The historian must establish the strange phenomenon: for the first time in history, not the upper class, but the lower class is a factor in the creation of a

Fig. 120. ROB. MALLET-STEVENS, PARIS 1925. BUREAU DU TOURISME. Decorative Arts Exhibition
The entire longitudinal building is suspended on two ferroconcrete uprights. The Dutch influence is evident in the overall modeling, above all in the cantilevered concrete slabs.

style. For the first part of the movement—around 1900—it was the white-marbled Stoclet house, this banker's fairy tale for which no expense was spared, that carried development, historical value. But this much can be said with certainty: today no building conceived in luxury and unlimited expense can any longer have any importance in the history of architecture.

Fig. 121. MALLET-STEVENS. Unfinished Villa for [Paul] Poiret

Fig. 122. MALLET-STEVENS. From the RUE MALLET-STEVENS. Paris 1927
Garage and staff on ground floor, left. MAIN ENTRANCE, right. Staircase in the center. The reception rooms begin on the first story. Note the comfortably arranged roof terraces above. The house serves as a single-family residence.
The traditional massiveness of the residence is not overcome. The various bodies (staircase, roof canopy, house cube) bump into one another without interpenetration. One sees how the exterior components of the new architecture (smooth surfaces, flat roof, lack of cornices, suspended elements) immediately disintegrate into mere details when the formal-representational building type still underlies the overall approach.

Henri SAUVAGE'S merit lies in having realized, in Paris, the *maison à gradins*, the recessed, terraced apartment house. These houses that recede story by story—and that are furthermore white-tiled—provide a source of light in the street. Even more than for the house built already in 1913 on the rue Vavin off the boulevard Raspail
Figs. 123, 124 is this true for the block of apartment houses on the rue des Amiraux much farther north. Built by the city of Paris. Inexpensive residences.
Since the block sits on three streets, the empty space beneath the receding stories could be used for an indoor swimming pool instead of a black light well. This method of suspending the construction of an entire block is possible only with the tradition of ferroconcrete.

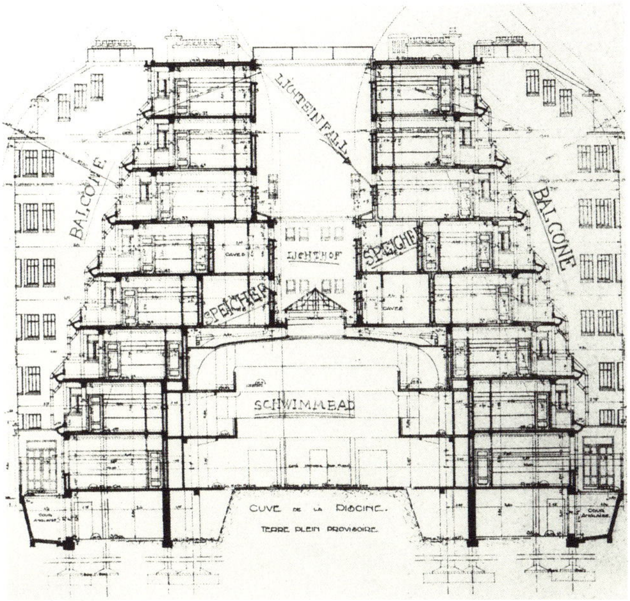

Fig. 123. HENRI SAUVAGE.
The Terraced Apartment House.
PARIS, RUE DES AMIRAUX 1925
Section.
Right and left are the recessed balconies. Swimming pool in the middle, above it storage space in the middle of the house.

Fig. 124. HENRI SAUVAGE.
RUE DES AMIRAUX 1925
The balconies are recessed in order to admit light.

Fig. 125. HENRI SAUVAGE: GARAGE for 1,000 Cars, PARIS 1926
Left: the gracefully ascending ramp connecting the various stories.

In his garage for a thousand cars off the boulevard Raspail (1925), Sauvage is probably more involved in the facade than in the actual engineering, and the same can be said about the other facades, which were kept completely in the Beaux-Arts style.

More important is the question of the real SUCCESSORS, of André Lurçat, [Gabriel] Guévrékian, R. Boyer-Gérente, J[ean-Charles] Moreux,[97] L[ouis] Thomas (of Lyon),[98] J. Beaugé.[99] One may say of all of them: they have no impact. One cannot as yet definitively separate the wheat from the chaff. France was a country of pensioners. As in other complacent countries the prevailing sentiment is: "*La peur du risque!*" The fear of the new. Even the backwardness with respect to obvious hygienic demands (bath, shower, boiler) is not simply the product of economic weakness but equally the power of a tyrannical tradition that has become a specter that does not in any way want to change living habits.

97) On Jean-Charles Moreux, who is still too indebted to French applied arts, cf. the illustrated essay in *Art et decoration* (February 1928), pp. 40ff. Jean Porcher publishes photographs of "*la maison nouvelle en France*" in the issue of January 1928 of the same journal.
98) A design for a multifamily house with roof gardens appears in volume 7 of *Manomètre,* edited in Lyons by [Emile] Malespine.
99) PLEYEL HALL (Paris, 1927) must also be mentioned. A concert hall for 2,600 people by the architects J[ean-] M[arcel] Auburtin, A[ndré] Granet, J. B. Mathon. The hall is based upon GUSTAVE LYON's principles of acoustic design. The enormous parabolic vaulting, ranging from the podium to the highest seats, simply allows no opportunity to add "STYLISTIC DETAILS."
Such examples show where the path leads. Even architects who have no feeling for the principles of the new architecture will simply be forced to follow the path consciously chosen by the avant-garde, for the functional aspects of every new building task must be satisfied. It is well known that Corbusier based the Assembly Hall for the League of Nations on Lyon's ideas. Excellent publication in the issue of *L'Architecte* from January 1928.

Fig. 126.
G. GUÉVRÉKIAN.
DESIGN FOR A FERRO-CONCRETE VILLA 1923
Earlier attempt to eliminate the massiveness of the house and to draw the consequences of the possibilities of ferroconcrete.

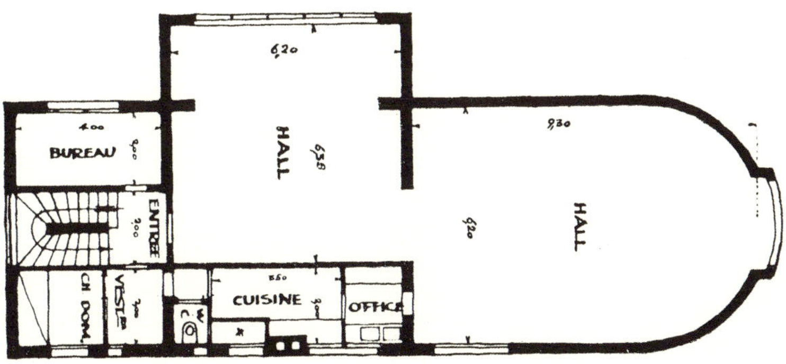

Fig. 127.
G. GUÉVRÉKIAN.
DESIGN FOR A FERRO-CONCRETE VILLA

Fig. 128. ANDRÉ LURÇAT. CITÉ SEURAT, PARIS 1925/26.
GARDEN FACADE OF A HOUSE

This house acquires an original appearance when one understands that the charmingly diverging lines of the first and second stories are to accommodate a mother and daughter who wanted to live as separately as possible. Similarly, the windows turn their backs on one another.

Fig. 129. ANDRÉ LURÇAT 1927.
Exercise Area on the Roof of the Guggenbuhl House in PARIS, PARK MONTSOURIS

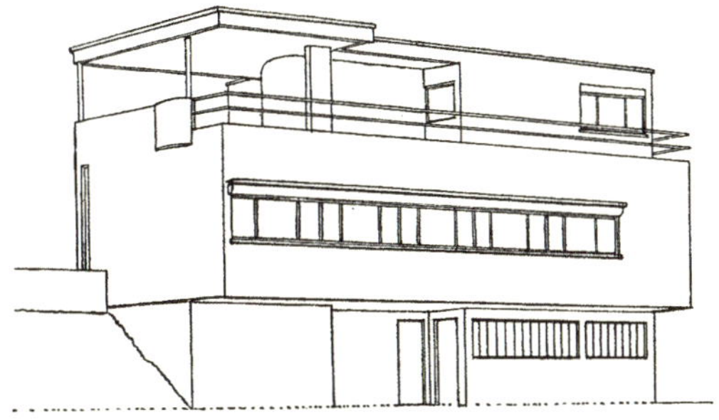

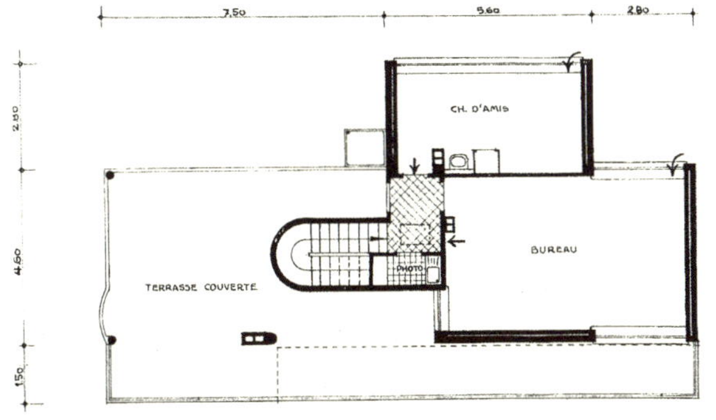

Figs. 130, 131.
R. BOYER-GÉRENTE.
Villa in HYÈRES (Riviera)

Gabriel GUÉVRÉKIAN, a Persian living in Paris, had a very interesting start. The ferroconcrete villa of 1923 is probably one of the earliest attempts to loosen the rigid cube of the house and interpenetrate it with air, to lay out three roof terraces, to lighten the ground floor, and to use an interior skeleton to suspend the building elements, without the trickery of decoratively cantilevered concrete slabs.

Guévrékian has yet to have any real impact. There are a few shops by him in Paris (such as the boutique Simultané, 1924, with its textile window displays moving at various speeds, or the Alban Photographic Studio, 1926). Guévrékian's executed gardens (at the Exhibition of Decorative Arts, 1925, and for the comte de Noailles in Hyères, 1925) do not bring us any closer to this difficult problem. With their virtually complete elimination of living plants they resemble a flirtation with abstract design more than a design itself. They are on a developmental sidetrack just as much as Mallet-Stevens's frequently cited concrete trees at the Paris exhibition, which were actually an unnecessary attempt to shock the Philistines. A large house in Neuilly is under construction.

The small, cubic studio of André LURÇAT (born 1894) is also known in Germany. His most significant achievements are the groups of houses of the Cité Seurat (1925/26), two houses in Versailles, and the Guggenbuhl house (Paris, Montsouris). His work is distinguished by a certain austerity, coolness. Whether by predisposition or by lack of understanding by the patrons, the imagination often seems restrained. Among the younger architects in France he is perhaps the one most concerned with sociological problems: hence the importance of his designs for apartment houses and his development plans related to them.

R. BOYER-GÉRENTE has followed in Corbusier's path with his design for the Ville de Curepipe by opening up the single house and striving for standardization. **Figs. 130, 131** (Published by van Doesburg in *Bouwbedrijf*, 1927, p. 88.) His villa presently under construction in Hyères underscores this impression.

Fig. 132. J. BEAUGÉ, 1926.
Housing Settlement for Intellectual Workers
Size: 50 × 80 meters. Nevertheless, a substantial area is reserved for collective games. The terrace gardens—facilitating broader vistas—emerge without additional expense by employing the surplus material of a neighboring building. Individual offices, beneficially removed from the residence, are placed at the top.

The young J. BEAUGÉ, from Corbusier's studio, displays a sympathetic grasp of the layout of the terrain and interpenetration in his design for a *cité jardin* for intellectual workers (exhibited in the Autumn Salon of 1926). Despite the essentially still-undigested influence of his master, one sees in him a hope and the possibility of further development.

Unlike Holland, the contemporary meaning of the new French architecture does not lie in the transformation of the whole country through a housing form for our age. Whereas in Holland the results rest mostly on the refinement of craftsmanship and less on industrialization, the way of looking at the problem in France is far more interesting because it is constantly directed toward employing ferroconcrete. Ferroconcrete buildings of the most elegant and bold construction are spreading all over France. Utility buildings. Upturn is enormous. And general. But only with regard to utility buildings and—with a certain reservation—"public works."
"Public works"—*travaux publics*—are an old Saint-Simonist issue, which both Napoleon III and the republic took over.
It is different with residential buildings. According to a discussion in the French parliament of 13 November 1927, France lacks approximately

ONE MILLION DWELLINGS.

Any traveler who comes to the outskirts of the city—*banlieue*—can vouch for this housing distress. One is astounded that in Lyons, for instance, abutting Garnier's vast slaughterhouse complex, there are inhabited wooden barracks whose damaged parts

are patched together from tin-can lids and crate slats. In contrast to Holland, the housing problem in France is making almost no progress. In the long run this will force the avant-garde into an utterly untenable position: into an aestheticism.
The public hand provides no commissions. A few painters and wealthy people—who can be counted on the fingers of one hand—are the only patrons. It is certainly good for experimental buildings not to be bound by financial considerations but only to the extent that it is possible to apply the collected experiences a hundred times over in the future. Without slow, industrially bound construction there can no longer be a lasting solution; without a collective will there can be no architecture.

Now, the problem in France is this: if the people are sufficiently driven to feel the need also to give their inner life an appropriate form, our age will instantly get a foothold there. As long as the masses are satisfied with the present bad living conditions, then the new architectural knowledge gained in France will only be beneficial elsewhere. The avant-garde, which in France encounters the most formidable resistance whenever a new housing form is attempted, is small but confident. One of its representatives wrote me: "Our hopes are vast, for we expect to win the game some day, and win it decisively. There are very few of us committed architects. But we have such a will to succeed that our ideas will slowly but surely make their way. It is only a matter of time."
The backbone of the young people is still artificially broken in the schools, and the ideal of the Académie des Beaux-Arts survives in the minds of the bourgeoisie. Blinded by grotesque ignorance, this institution still pretends to be the "guardian of a grand tradition." Its shameful influence on all official milieus is undeniable. With the competition for the League of Nations building we have observed with amazement how much apparent power such an institution can still have, even though for a century it has been without significance for the history of art.
But, after all, our age is a barren soil for such specters. Sooner or later they must decide to expire, even in France.

THE CURRENT STATUS OF FERROCONCRETE

Fig. 133. Limousin & Co. (FREYSSINET Technique). SAINT-PIERRE-DU-VAUVRAY Bridge 1922
A SPAN of 132 meters is bridged by a single arch. The widest span executed so far. Conspicuously lithe form. This is achieved by thorough consideration of the material's properties, for example, the verticals are of iron, for this exploits the cable tension. Cf. *Le Génie Civil*, 3 November 1923. Freyssinet's bridge near BREST, currently under construction, has three arches, each spanning 190 meters. Material: ferroconcrete.

In the three decades of its most widespread application, ferroconcrete has reached such a high level of design that fewer fundamentally new results are to be expected—but rather an extension and refinement of practical methods.
It is curious that ferroconcrete has needed about the same length of time as iron for its development. Iron between 1850 and 1890, ferroconcrete from the beginning of the 1890s.
Likewise, ferroconcrete attempts to break up the rigid relationship of support and load. This is especially visible in so-called mushroom slabs, where supports are allowed to flow directly into the ceiling plane.
With ferroconcrete there is today also the possibility of anchoring buildings high above the ground and letting their weight be suspended from a supporting arch. Be it the arch of a bridge, a garage, or a hangar.
Just as Eiffel pursued the problem of the parabolic support his whole life, so the engineer Freyssinet has—since 1916—made attempts in ferroconcrete to suspend loads

Fig. 134. Limousin & Co.
PONT DE VILLENEUVE-SUR-LOT 1914–1919. SPAN 96.25 meters
The arch is made of unreinforced concrete. Compare the previous bridge and its formal litheness, achieved through the use of ferroconcrete.

from above and has in this way arrived at an until-now unknown lightness, grace, and span (pont de Saint-Pierre-du-Vauvray, 132 meters). One always sees that same balance of an arithmetic and aesthetic solution. Fig. 133

Limousin & Co. explain in the catalog of their constructions: "Ferroconcrete is no longer a mystical dish for which everyone unfailingly has a better recipe than his neighbor. Constructions are based on a number of well-tested methods. What counts today is the perfection of production."

Next to the creator of forms (*créateur des formes*) a second corrective authority emerges from handling of such a complicated material as ferroconcrete: the entrepreneur, the organizer!

Fig. 135. Limousin & Co. (Freyssinet Technique) AIRPLANE HANGARS in CHARTRES 1924
The weight of the hall is suspended from above. The same design elements in bridges and building construction.

Fig. 136. BRICE & SAINRAPT. VAULT (TOUR FORTE) of the BANQUE DE FRANCE
Ten stories. Underground. Height 32 meters. Diameter 50 meters. Closure by means of movable concrete wings.

He sees to it that manual labor is reduced to the absolute minimum and is concerned to design the form in such a way that the frameworks may be repeated and used as often as possible; at the same time, for reasons of thrift, he requires their quick and easy assembly and disassembly. This leads to a new inner lawfulness, to a new as-yet-unknown expression.

Fig. 137. BRICE & SAINRAPT. TRIBUNE of the Club athlétique de la Société générale
3,000 seats. 100 meters long. CANOPY: 10-meter cantilever.

Fig. 138. BRICE & SAINRAPT. WINE STORAGE
30,000 HECTOLITERS. Firm of DAMOY in IVRY.

How is ferroconcrete related to HOUSING CONSTRUCTION?
It is France's special role that, with her pleasure in overcoming constructional difficulties, she has devised almost all buildings in ferroconcrete, taking into account the new housing functions.
But this does not resolve the questions: there is a certain inherent rigidity in ferroconcrete. We know how difficult it is to demolish or even just alter buildings in this material. Our dwellings should not be rigid, they must be allowed a certain amount of free play for change, for they do not always serve the same function. Because of the rigid nature of today's house, we have been all too accepting of the immutability of the floor plan and have let it tyrannize us.
Iron skeletons are after all more open to reassembling or addition, yet they display other disadvantages. Gradually, experience accumulates that allows us to determine with confidence which material can best be applied in a specific area of a house.
The beginning of industrialization and transportation in the nineteenth century led to the formation of iron. Later, as the world economy recognized its importance, it demanded containers for large-scale storage: warehouses and silos. This demand produced the suitable building material: FERROCONCRETE.
In the field of building, our age has one primary demand: the creation of a humane and unconfined human dwelling that meets minimum standards [*Existenzminimum*].

Fig. 139. ROOF OF THE RUE MARBEUF GARAGE, Paris 1926
The broad surfaces at the top level of the large cities will be used for sports, gardening, and eventually probably also for landing strips for aircraft. For hygienic reasons alone, steep roofs will soon be forbidden in large cities. Flat ROOFS will serve recreational purposes and offer points of rest for the eye accustomed to today's disruptions.

Everyone now knows that a substantial price reduction is possible only through the industrial organization of the fragmented building process. No city in the future can do without experimental laboratories for practical building, for one must determine for each stretch of land which building method, or better, which combination of building materials is optimally suitable for it.
In addition, it is to be hoped that our age will form a new BUILDING MATERIAL that comes closer to its demands—a house that meets minimum standards.

Just as the nineteenth century—at a given moment—developed iron and ferroconcrete for its needs, so we can assume that our age, too, will find the material that responds to its demands.

Editor's Notes

1. See Introduction page 49 with note 161 on the missing question mark after this head.

2. Giedion erroneously attributes the first name of John to the English architect and horticulturist Sir Joseph Paxton (1801–1865), who designed the Crystal Palace for the London Industrial Exhibition of 1851.

3. Gaspard Monge, comte de Pélouse (1746–1818); Joseph-Louis Lagrange (1736–1813).

4. Giedion is referring to Jean-François-Joseph Lecointe (1783–1858), a partner in practice with Jacques-Ignace Hittorff.

5. See also Giedion's note 17 on page 103.

6. Louis-Auguste Boileau, *Nouvelle forme architecturale* (Paris: Gide & J. Baudry, 1853); idem, *Histoire critique de l'invention en architecture...la composition de nouveaux types architectoniques dérivant de l'usage de fer* (Paris: Vve C. Dunod, 1886).

7. The correct name is Victor Contamin.

8. Although Giedion mentions both the Perrets in the previous paragraph and thus is clearly aware that there are two brothers, he seems in this section to be talking only about Auguste Perret, combining the initials of both brothers in the section head and captions, as did Paul Jamot on the cover of his book about the Perrets.

9. Garnier's slaughterhouse was in fact begun in 1908 and completed in 1924; the Grange-Blanche Hospital was erected between 1911 and 1927; and the stadium was built 1913–1918.

10. Mallet-Stevens created the first designs for his *Cité moderne* in 1917. By 1921 the colors in these designs had faded to the point where they could not be reproduced, so Mallet-Stevens redrew them in order that they might be published in an album by Charles Massin. That album, with thirty-two plates, appeared in a limited edition in 1922.

Appendix

The Book as a *Gesamtkunstwerk*

"Only film can make the new architecture intelligible! Still photography does not capture it clearly," notes Sigfried Giedion in a passage in *Bauen in Frankreich*. This attitude is elucidated by the visual material that makes up Giedion's book, in particular the photographs from his own camera. In the pictures of the Eiffel Tower, the *pont transbordeur,* or even the Pessac settlement, Giedion applies a series of photographic techniques that reduce forms to lines, surfaces, and volumes, and that reach such a degree of abstraction that the observer is seduced into abandoning the conventional "perspectival" view and adopting a cinematographic view. Through close camera work, the particular thematic object appears as a fragment or a segment; through the intensive exploitation of light and shadow attained by appropriate exposure, the object is defamiliarized and dematerialized, its geometry distorted by the camera angles, and the often neutral background makes it difficult to judge the scale. Frontal views are avoided; up and down, right and left are freely interchangeable. Giedion often neglects to establish the building's tectonic contact with the ground, just as he neglects the relationship to the near and far surroundings. One might say that Giedion strives to tear up the set frame of the picture, to replace the individual picture as a source of information with a sequence of shots that corresponds to a perception in movement.

We may also assume that Giedion found the limitations of the book as a medium too rigid, too inflexible, and ultimately too tyrannical for the "grasping" of the new architecture. This is immediately visible in the text in the rapid changes from prosaic, descriptive, or analytical segments to others that are less concerned with the factual. Themes are only sketchily touched upon and then yield to the pictures by means of remarkably expressive, almost poetic words. The picture naturally has absolute priority. It arrests the thought and is crucial to the book's composition. Giedion stages the book through the pictures.

The author's part in the book's design—which, as noted in the Preliminary Remark, was undertaken in collaboration with Moholy-Nagy—is clear from a series of extant layout sketches in Giedion's own hand, which shed a very special light on the genesis of *Bauen in Frankreich*. The basis of these sketches is not the single page, but each double page. The pictures are the protagonists of the action, and here again it is not the single image that is crucial but rather the composition of pictures: their varying sizes and free distribution on the page. Thus there arises first of all a dialog of pictures. The bold headlines in capital letters explain what the dialog is about. The text of the dialog, which appears in the form of picture captions that are generally designed as short, concise explanatory messages, was placed on the paper at the same time as or immediately after the pictures were attached. One gets the feeling that the captions flow from the pictures. Picture and text form an inseparable whole. The process of thought unfolds through the interaction of picture and language. Expressed differently: thoughts are formed from the pictures, thoughts are projected onto pictures.

This method is supposed to speak to the "hurried reader," who is meant to "understand the developmental path from the captioned illustrations." The layout design clearly shows the special care lavished on the picture, text, and production, and precisely this concern leaves no doubt that Giedion consciously aimed his text at this reader. The "hurried reader" is simply the modern reader. It is for this reader above all that the book arising from these preliminary sketches is made.

The main text serves a useful function by justifying in more detail what can be seen in the pictures. Through the free form of its flow and through its deliberate inhomogeneity, which is revealed in the mixture of apodictic, argumentative, and expressive elements, the text is in keeping with the character conveyed by the visual impression. In the visual design of the book, however, the text is largely ignored or even entirely removed. At least it is not present in the layout sketches. The book is nonetheless made with the totalizing attitude of a *Gesamtkunstwerk*. It is meant to seduce the uneducated, to convince the dissident that his position is untenable, and to provide the initiated with arguments. Giedion proceeds like a strategist—completely confident about his own mission, and leaving nothing to chance.

All the extant design boards with Giedion's handwritten notes and instructions are reproduced on the following pages.

— S. Georgiadis

[illegible]
50 % [illegible]

19

FRANZÖSISCHE KONSTRUKTION

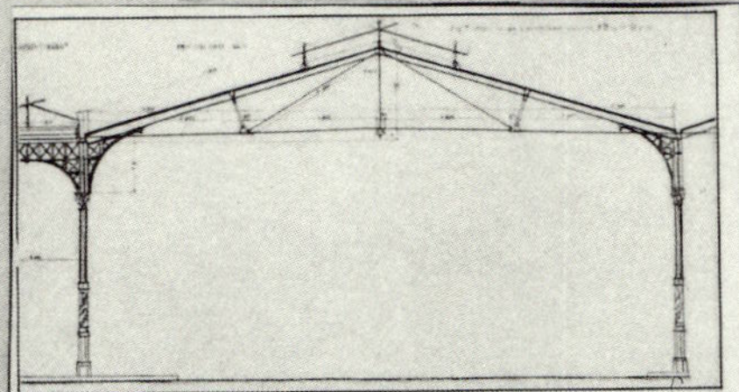

Abb. 13
POLONCEAU-BINDER
vom Gare du
Nord Einfahrtshalle
ET (1862)

[illegible] …, gelegentlich [illegible] zeigt bereits ein Eingehen auf die innere Struktur des Materials Eisen.
FOLGE: Leichtheit der Konstruktion, Vermeidung jeden Horizontalschubs. Drückt nur mit Eigengewicht auf Unterlage, daher dünne Stützen, große Übersichtlichkeit ~~Für hohe Bauten~~ der Räume.
Die sichtbaren Zugstangen sind noch lästig. Man wird versuchen, sie zu eliminieren. Sobald das konstruktiv gelungen ist (vgl. Abb. –) steht man vor neuen ~~Gestalten~~, vorbildlosen Gestaltungen

2

ENGLISCHE KONSTRUKTION

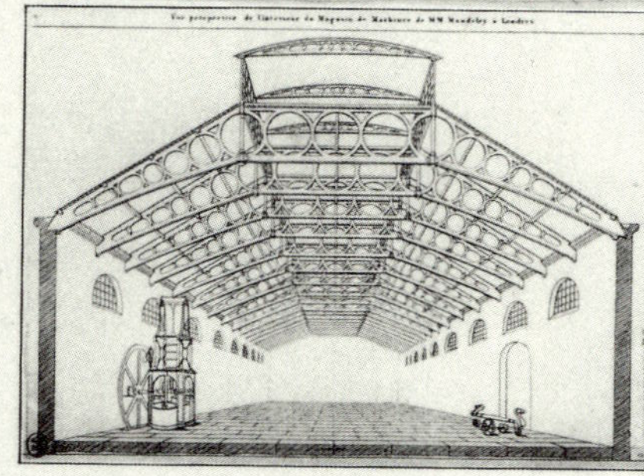

Abb. 14
ENGLISCHE
MASCHINEN
HALLE 183[illegible]

Direkte Übernahme des hölzernen Dachstuhls in gusseisernem Material. FOLGE: Schwere der Konstruktion und Notwendigkeit dicker Mauern, um den Horizontalschub aufzufangen. Materiala

3

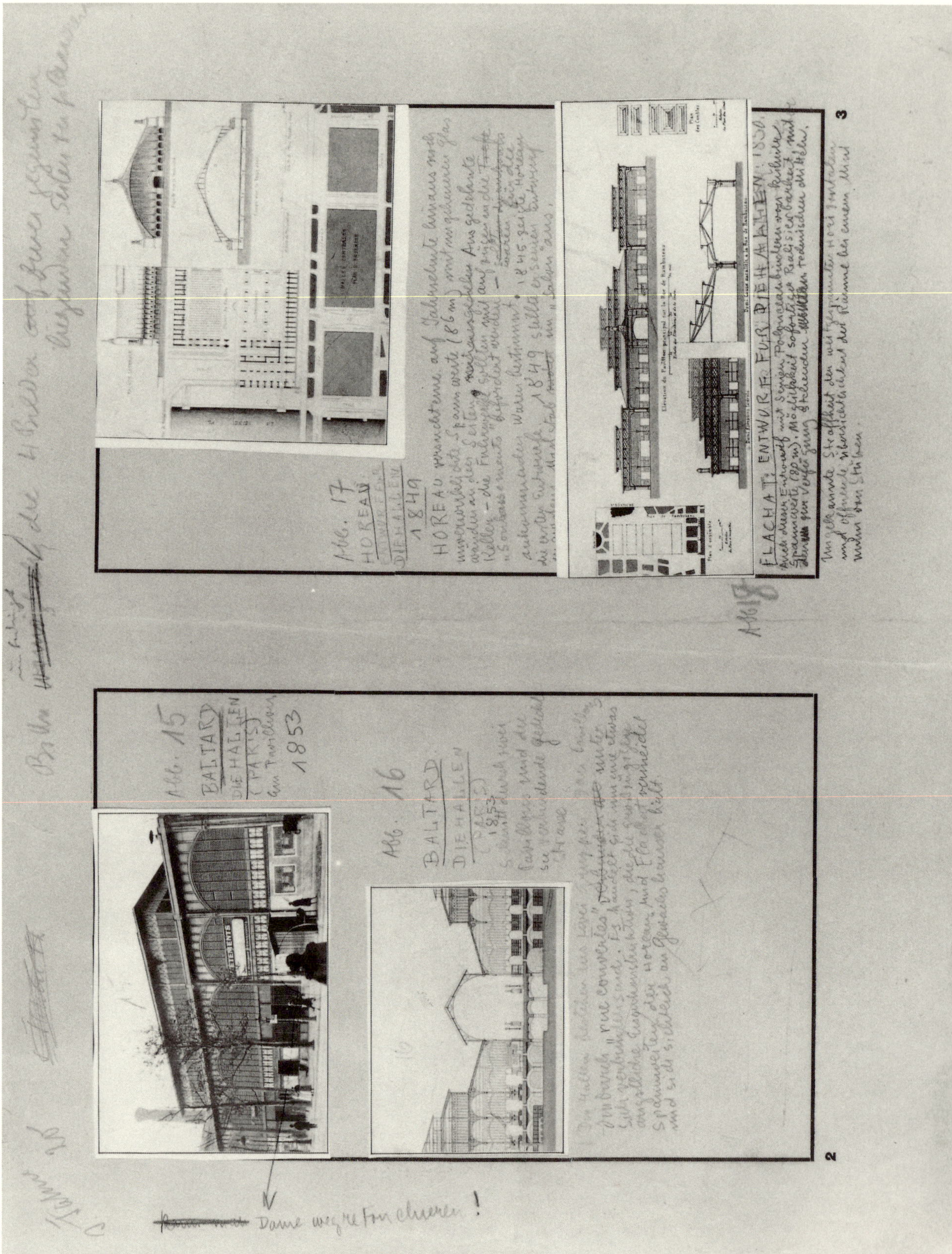
Abb. 15
BALTARD
DIE HALLEN
(PARIS)
1853
Abb. 16
BALTARD
DIE HALLEN
(PARIS)
1853
Abb. 17
HOREAU
DIE HALLEN
1849
FLACHAT: ENTWURF FÜR DIE HALLEN
2
3

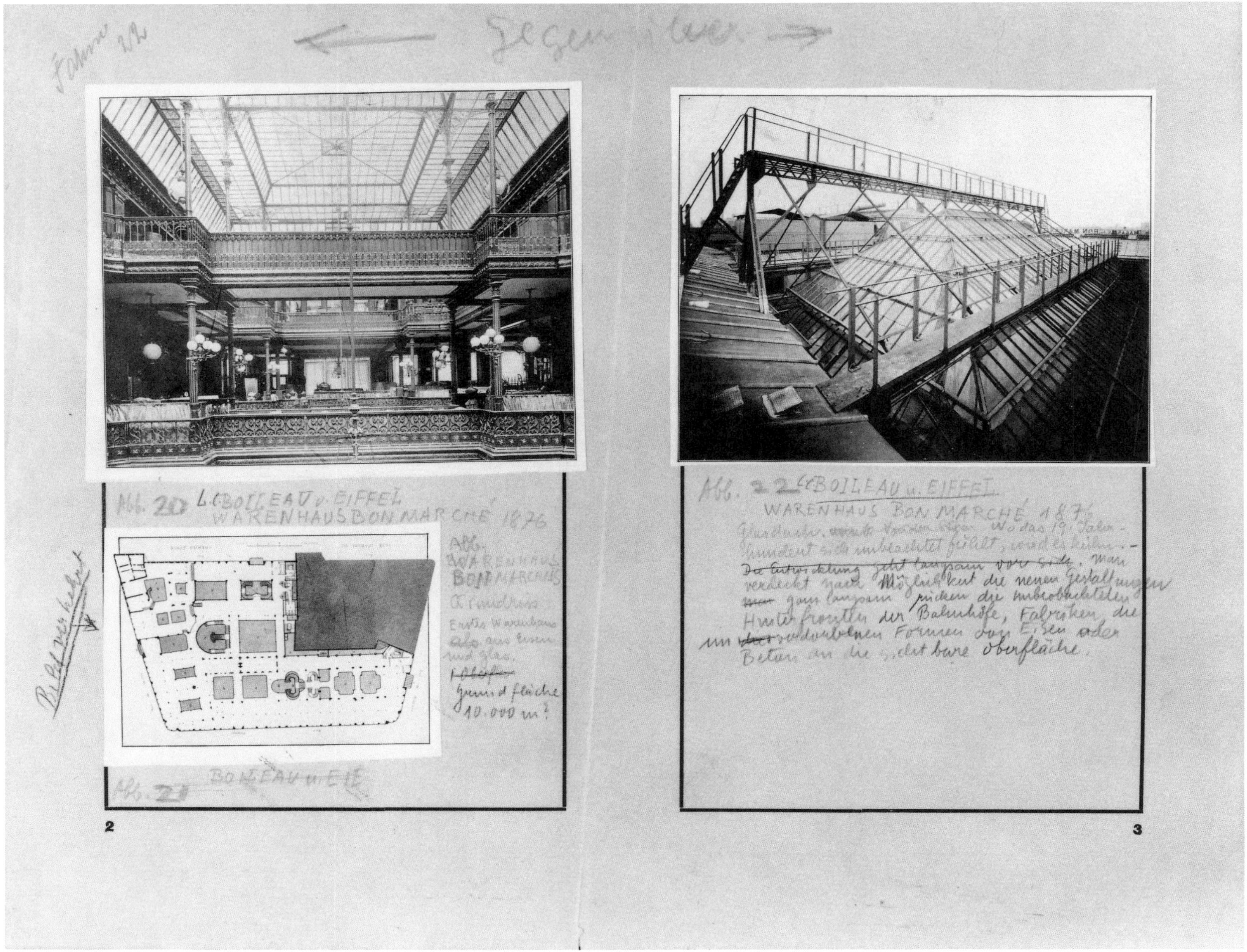
Gegenüber
Abb. 20 L.C. BOILEAU u. EIFFEL
WARENHAUS BON MARCHÉ 1876
Grundriss
Erstes Warenhaus aus Eisen und Glas.
Grundfläche 10.000 m²
Abb. 22 L.C. BOILEAU u. EIFFEL
WARENHAUS BON MARCHÉ 1876
Wo das 19. Jahrhundert sich unbeachtet fühlt, wird es kühn.
Hinterfronten der Bahnhöfe, Fabriken
Beton an die sichtbare Oberfläche.
2
3

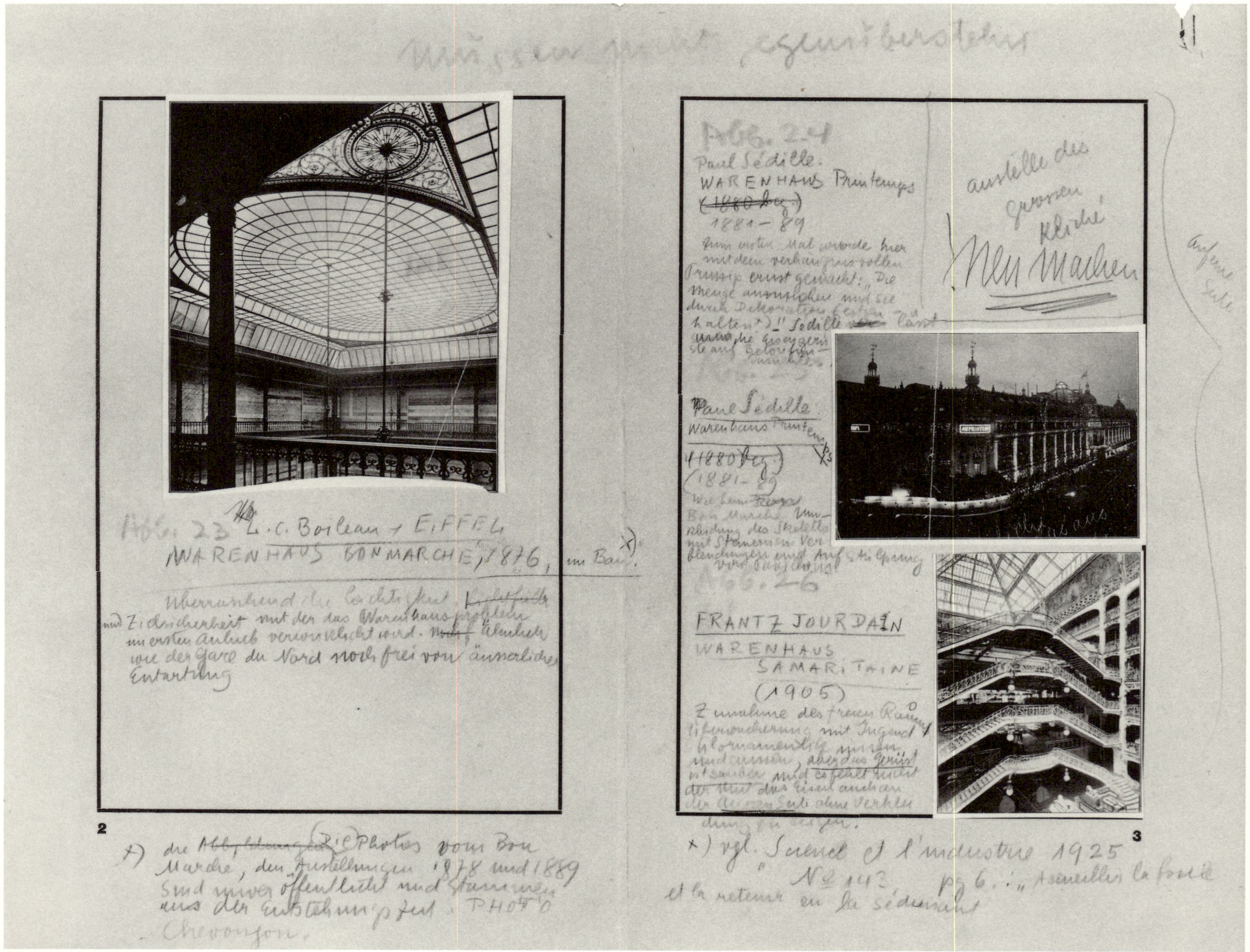
müssen genau gegenüberstehen
Abb. 23 L. C. Boileau + EIFFEL
WARENHAUS BON MARCHÉ, 1876, im Bau x)
Überraschend die Leichtigkeit und Zielsicherheit mit der das Warenhausproblem im ersten Anlauf verwirklicht wird. Ähnlich wie der Gare du Nord noch frei von äusserlicher Entartung
2
x) die Photos vom Bon Marché, den Ausstellungen 1878 und 1889 sind unveröffentlicht und stammen aus der Entstehungszeit. PHOTO Chevojon.
Abb. 24
Paul Sédille:
WARENHAUS Printemps
1881–89
Zum ersten Mal wurde hier mit dem Verkaufsprinzip ernst gemacht: „Die Menge anzulocken und sie durch Dekoration festzuhalten x)". Sédille lässt [illegible]
anstelle des grossen Klischés
Neu machen
Paul Sédille
Warenhaus Printemps
(1881–89)
Wie beim Bon Marché Umkleidung des Skeletts mit steinernen Verblendungen und Aufstülpung von [illegible]
Abb. 26
FRANTZ JOURDAIN
WARENHAUS SAMARITAINE
(1905)
Zunahme des freien Raumes [illegible] mit Jugendstilornamentik innen und aussen, aber das Gerüst ist sauber und es fehlt nicht der Mut das Eisen auch an der Aussenseite ohne Verkleidung zu zeigen.
äussere Seite
3
x) vgl. Science et l'industrie 1925 N° 143 pg 6: „Accueillir la foule et la retenir en la séduisant"

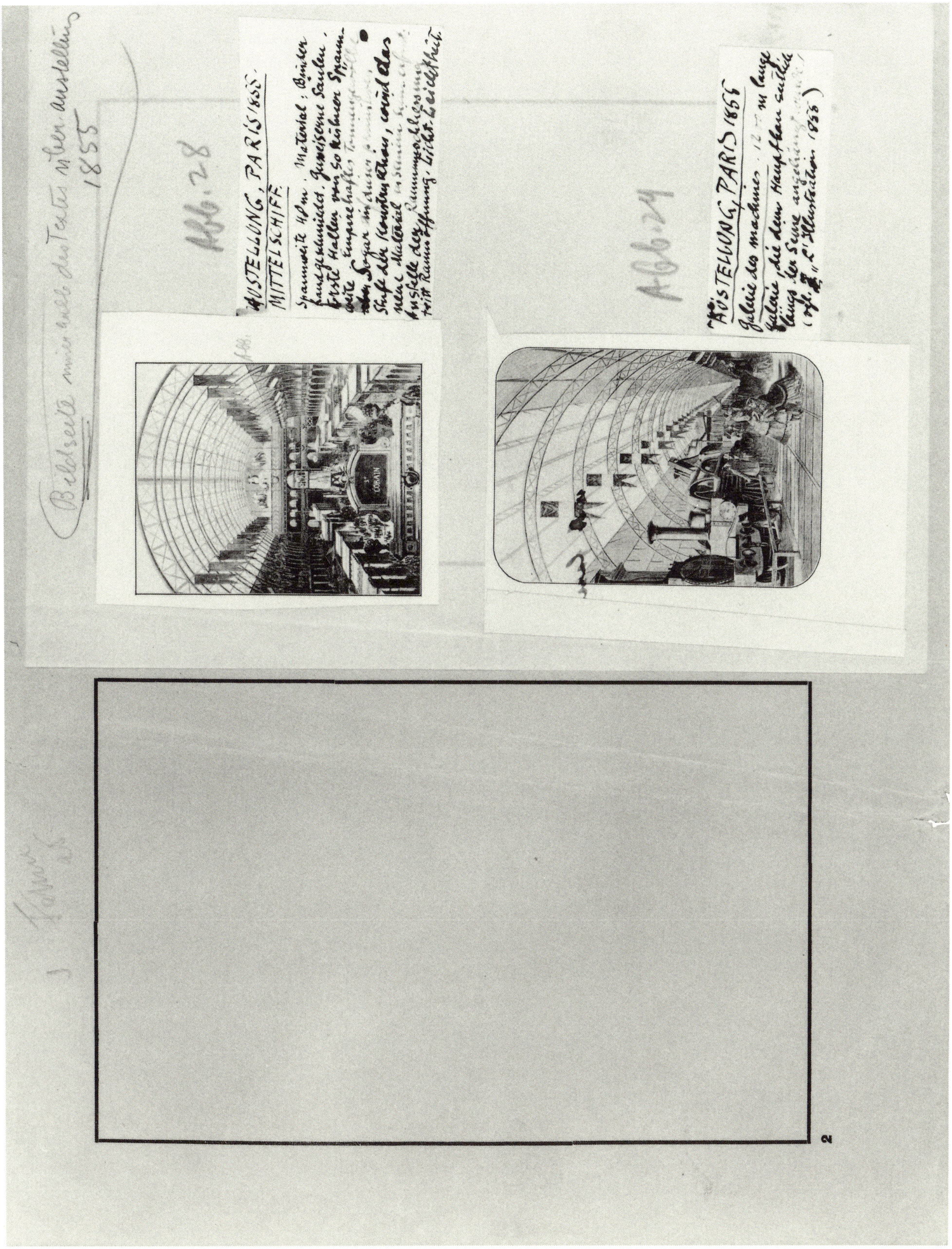
Bildseite innerhalb des Textes über Ausstellung 1855
Abb. 28
AUSTELLUNG, PARIS 1855.
MITTELSCHIFF.
Spannweite 48 m. Material: Binder handgeschmiedet. Gusseiserne Säulen. Erste Hallen von so kühner Spannweite.
Sogar in dieser primitiven Stufe der Konstruktion, wird das neue Material
Anstelle der Raumumschliessung tritt Raumöffnung. Licht. Leichtheit.
COBAIN
Abb. 24
AUSTELLUNG, PARIS 1855
Galerie des machines. 12… m lange Galerie, die dem Hauptbau seitlich längs der Seine angehängt wurde
(vgl. „L'Illustration 1855)
2

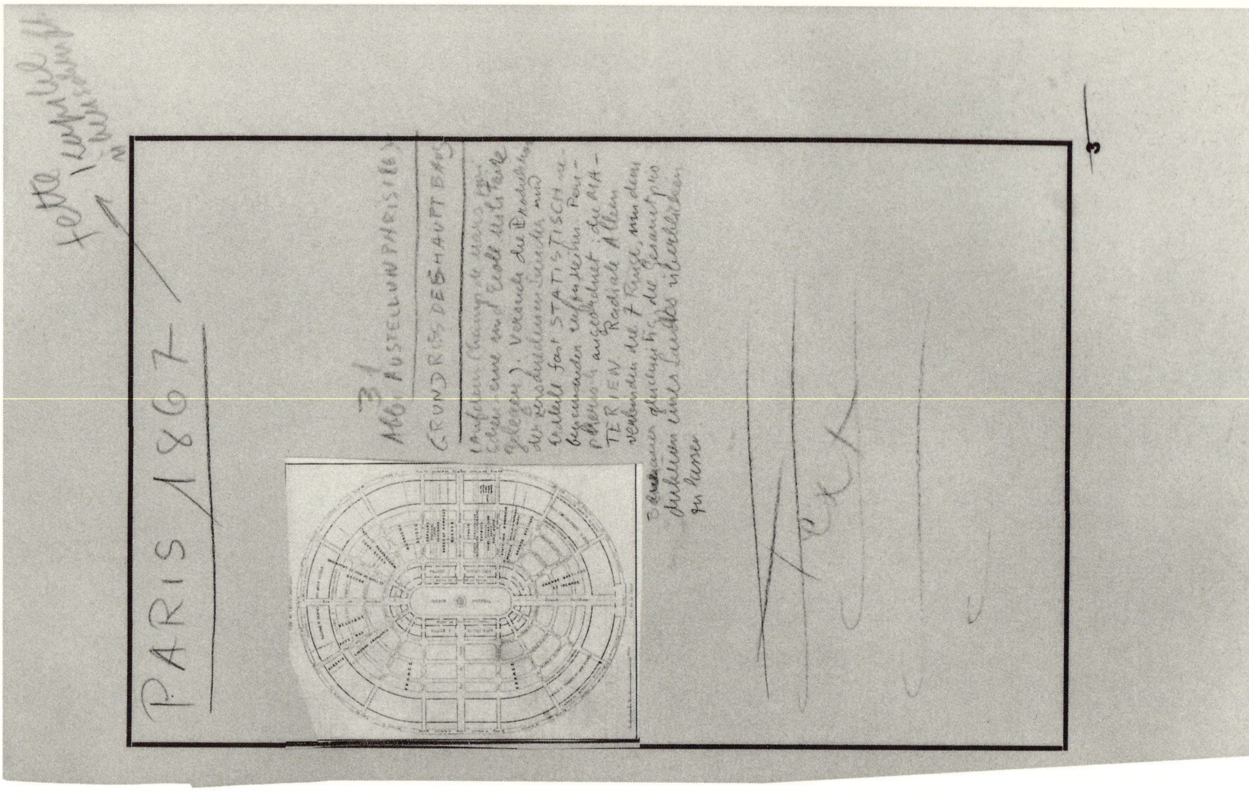
PARIS 1867

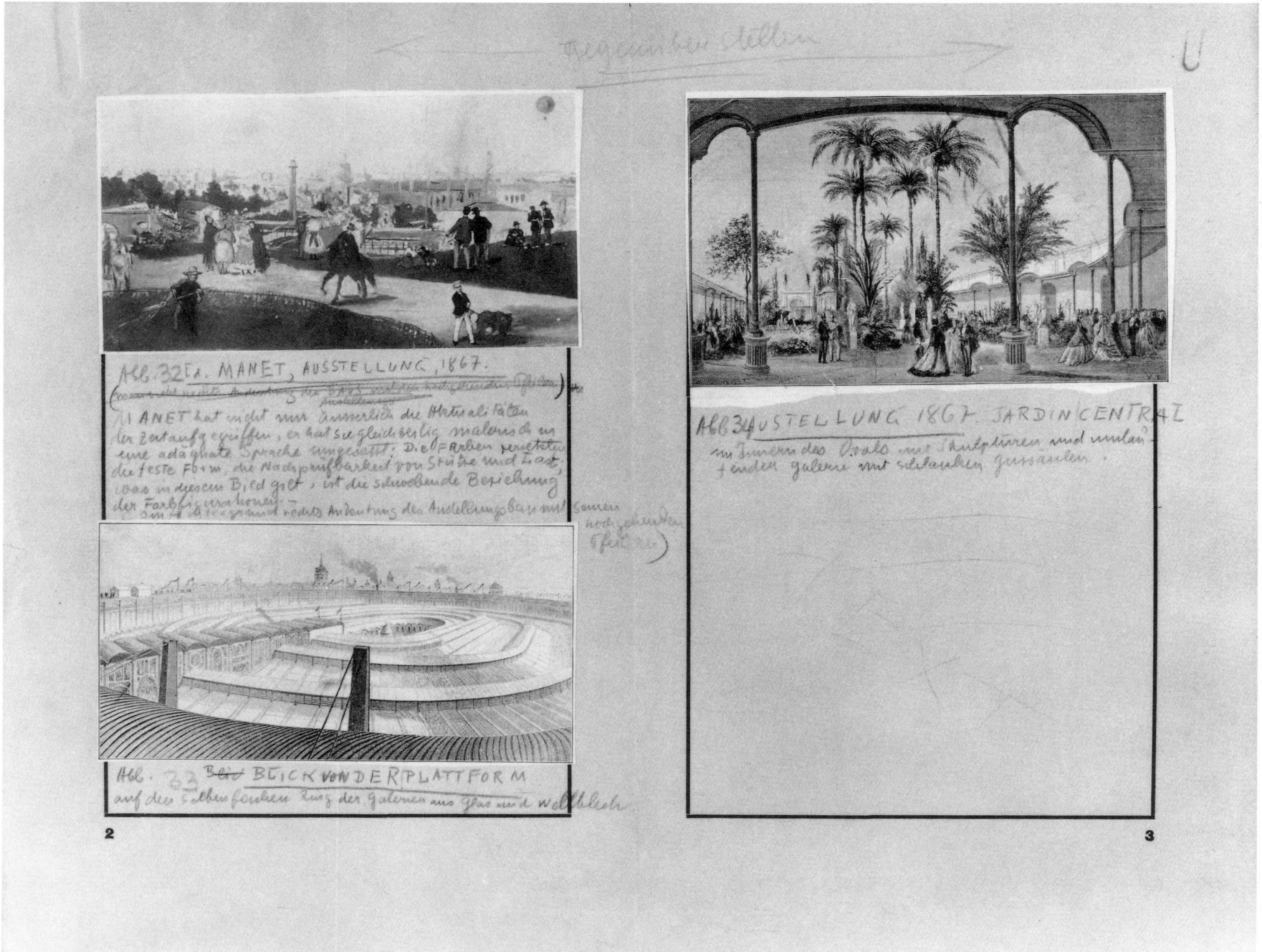
gegenüber stellen
U
Abb. 32 Ed. MANET, AUSSTELLUNG, 1867.
MANET hat nicht nur äusserlich die Aktualitäten der Zeit aufgegriffen, er hat sie gleichzeitig malerisch in eine adäquate Sprache umgesetzt; Die Farben zersetzen die feste Form, die Nachprüfbarkeit von Stütze und Last; was in diesem Bild gilt, ist die schwebende Beziehung der Farbfigurationen.–
im Hintergrund rechts Andeutung des Ausstellungsbaus mit seinen hochgehenden Pfeilern)
Abb. 33 BLICK VON DER PLATTFORM
Ring der Galerien aus Glas und Wellblech
2
Abb 34 AUSTELLUNG 1867. JARDIN CENTRAL
im Innern des Ovals mit Skulpturen und umlaufenden Galerie mit schlanken Gussäulen.
3

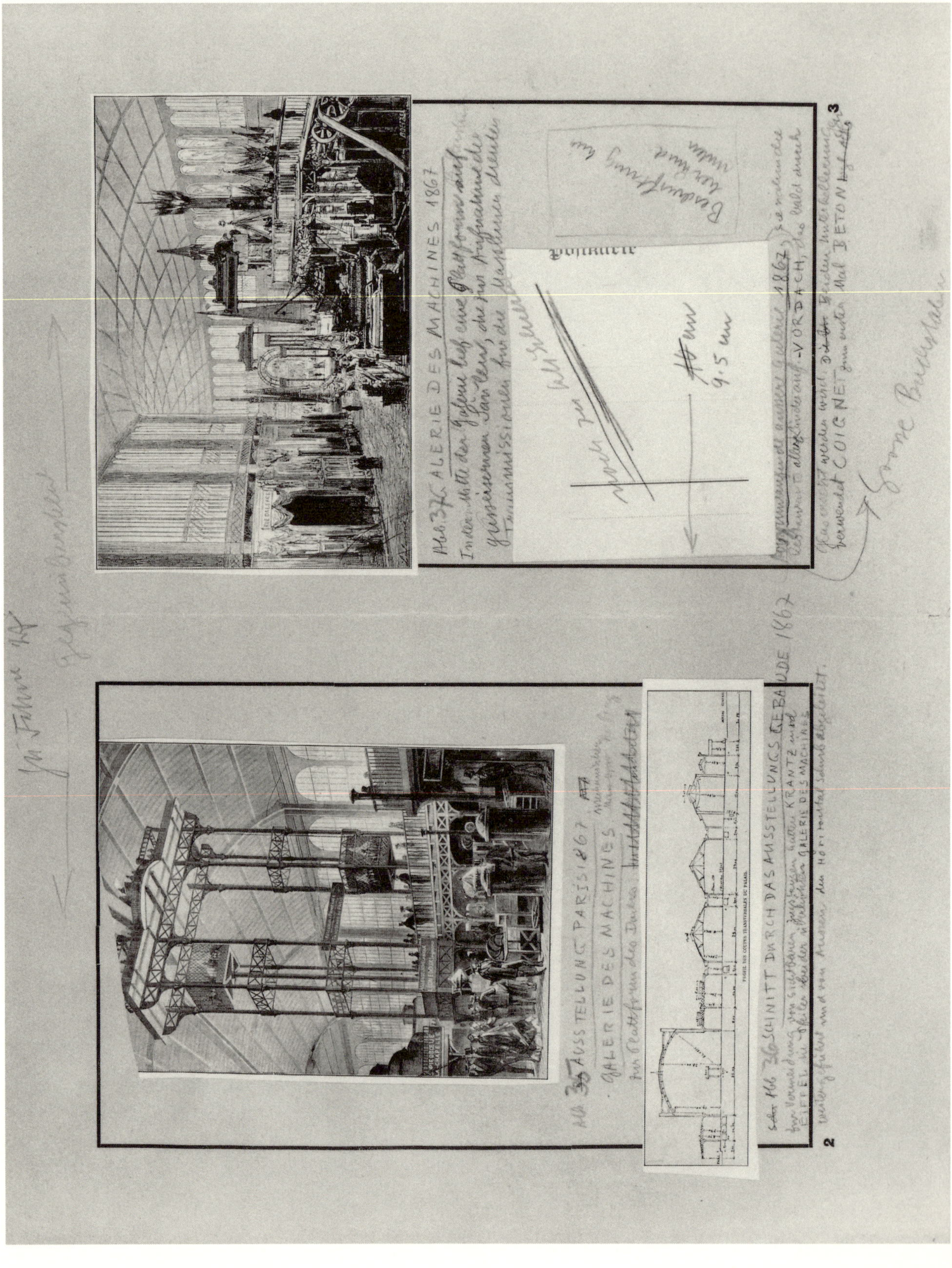
AUSSTELLUNG PARIS 1867
GALERIE DES MACHINES
GALERIE DES MACHINES 1867
SCHNITT DURCH DAS AUSSTELLUNGSGEBÄUDE 1867

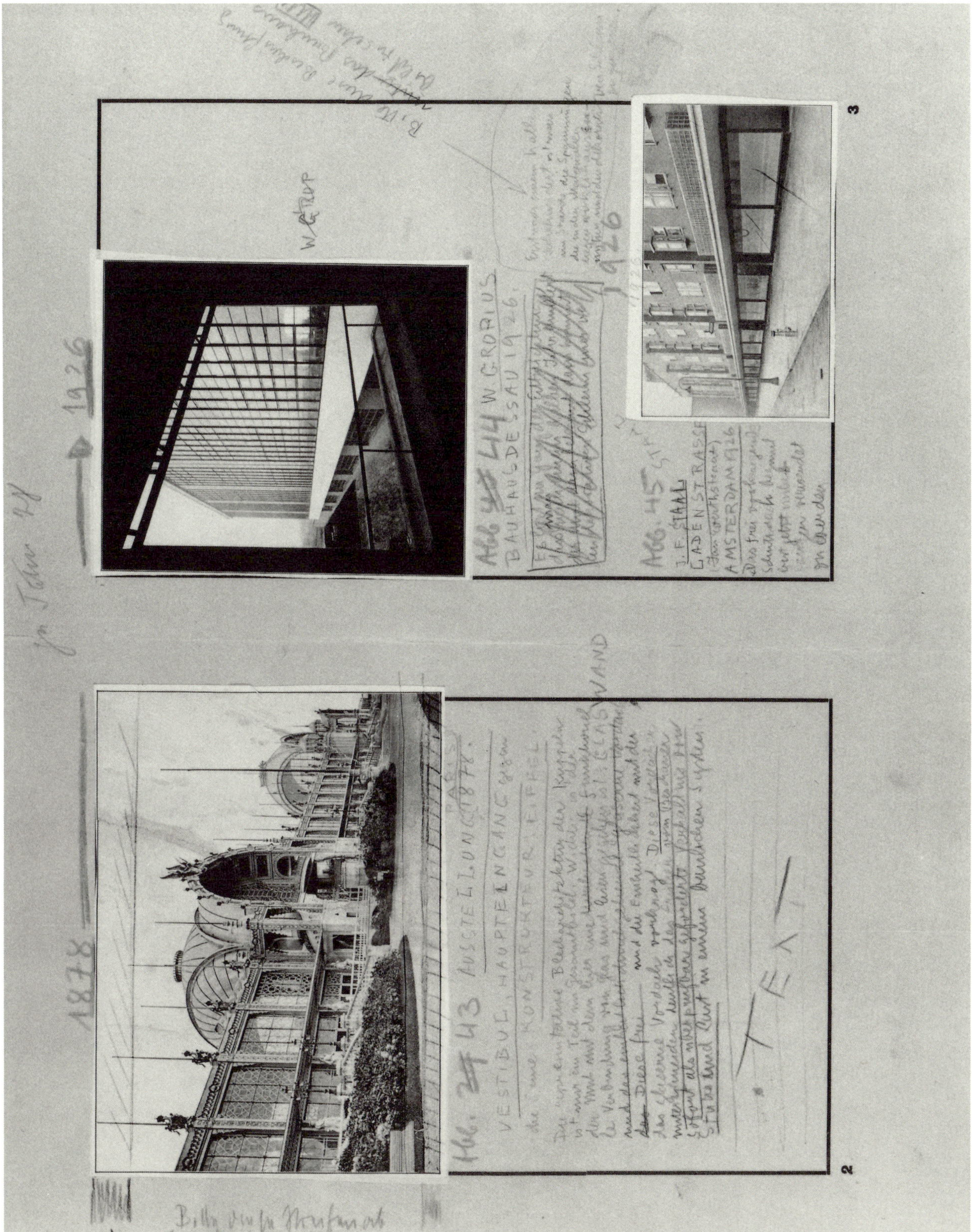
1878
1926
Abb. 43 AUSSTELLUNG 1878
VESTIBUL, HAUPTEINGANG
PARIS
Abb. 44 W. GROPIUS
BAUHAUS DESSAU 1926
Abb. 45
J. F. STAAL
LADENSTRASSE
AMSTERDAM 1926

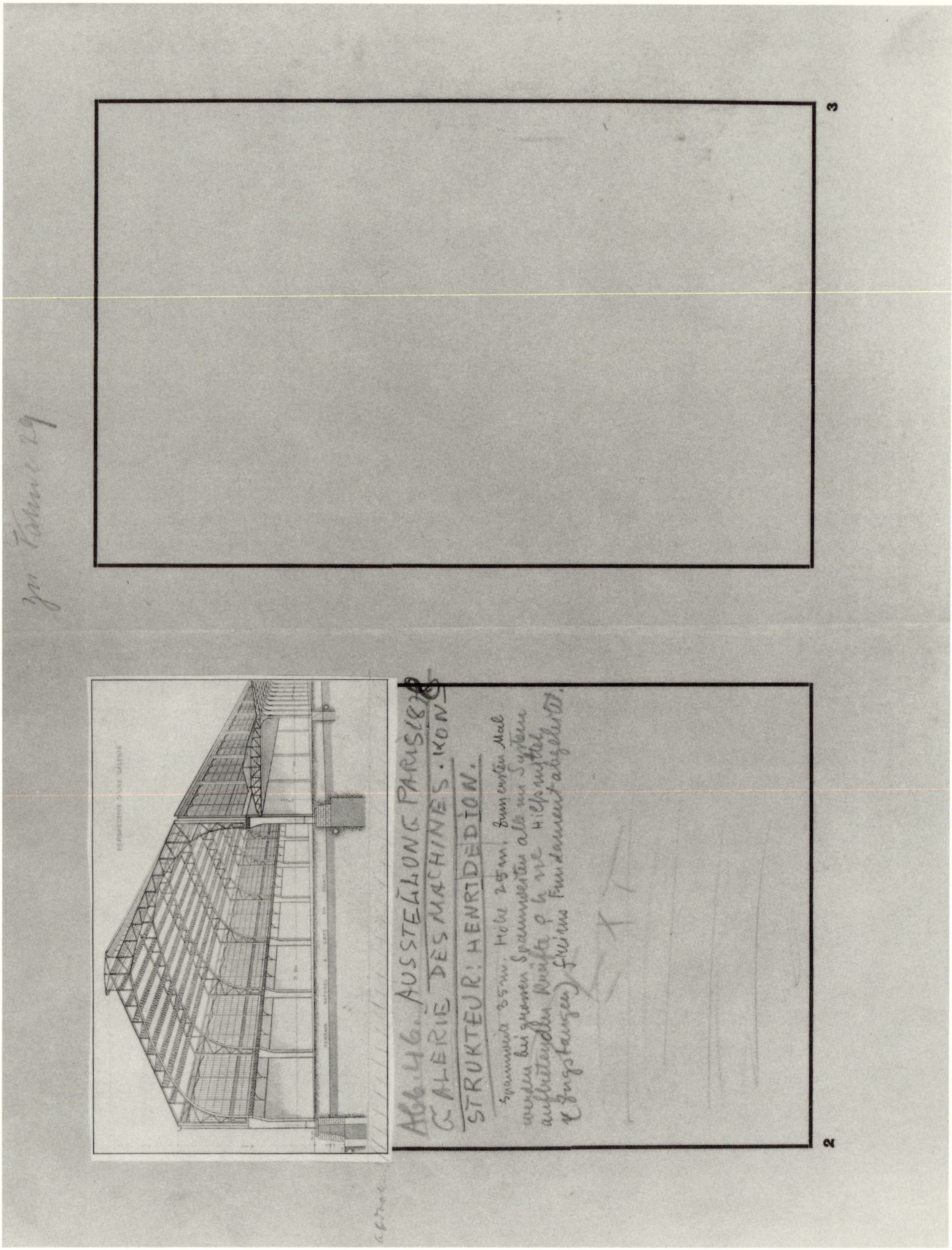
ABB. 46. AUSSTELLUNG PARIS
GALERIE DES MACHINES. KON-
STRUKTEUR: HENRI DE DION.
2
3

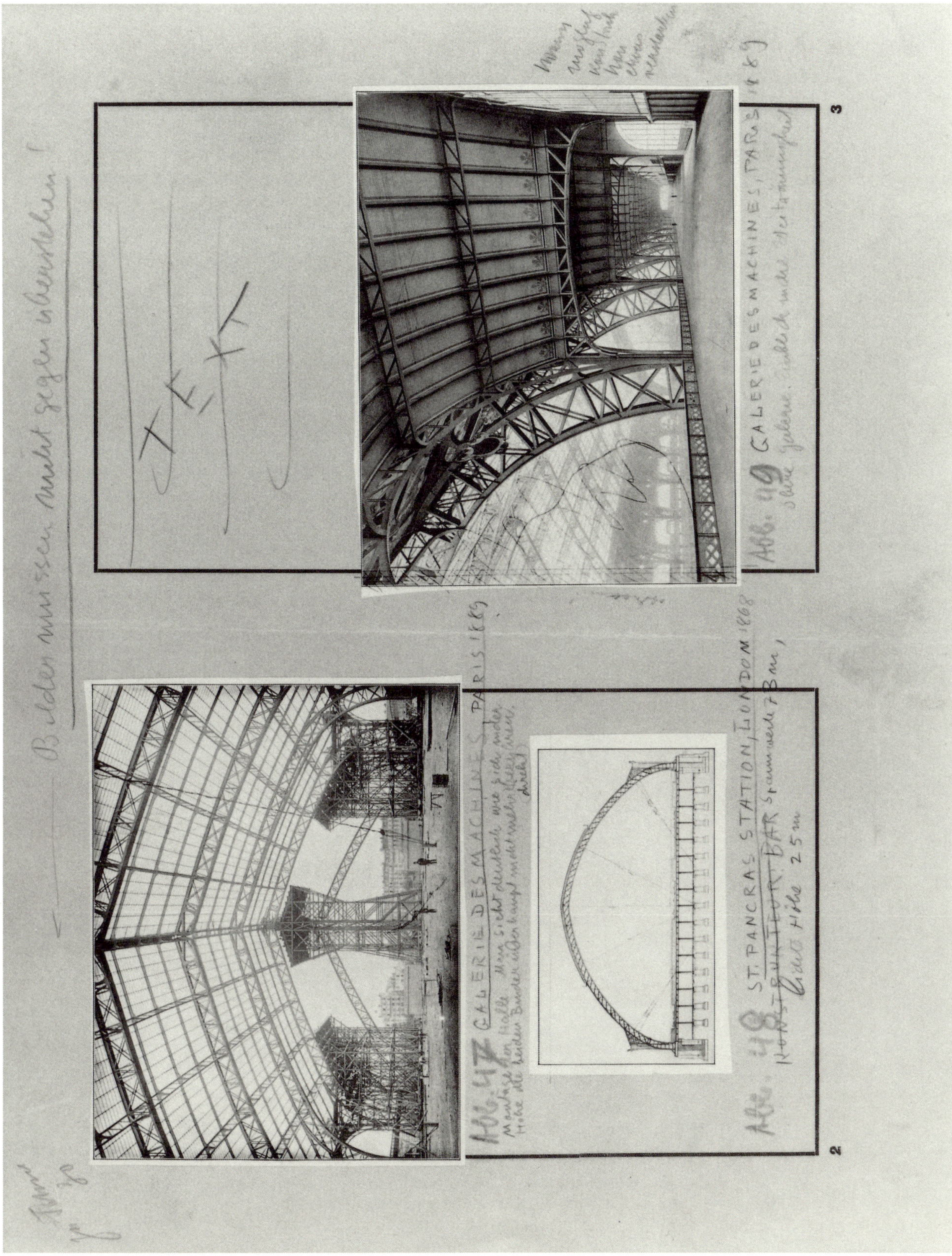

GALERIE DES MACHINES PARIS 1889

Ohne Basis und ohne starre Verbindung mit der Erde ruht die Last der ganzen Halle auf den kleinen Rollenlagern

2

3

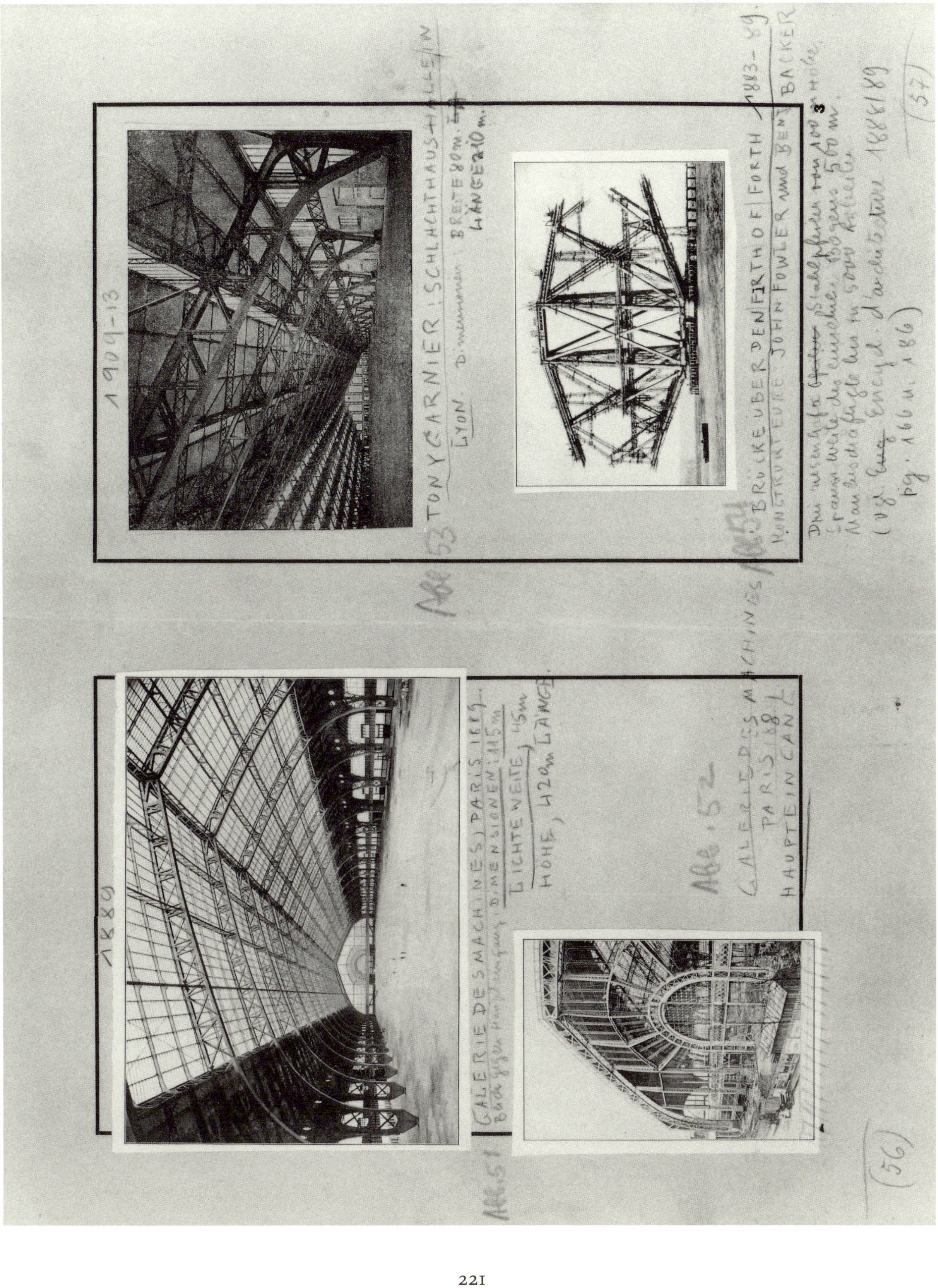
1909-13
TONY GARNIER: SCHLACHTHAUSHALLE IN LYON
Dimensionen: BREITE 80m. LÄNGE 210 m.
ABB. 53
BRÜCKE ÜBER DEN FIRTH OF FORTH 1883-89.
KONSTRUKTEURE JOHN FOWLER und BENJ. BACKER
(57)
1889
GALERIE DES MACHINES, PARIS 1889.
DIMENSIONEN: LICHTEWEITE 115m
HÖHE 45m, LÄNGE 420m
ABB. 51
ABB. 52
GALERIE DES MACHINES
PARIS 1889
HAUPTEINGANG
(56)

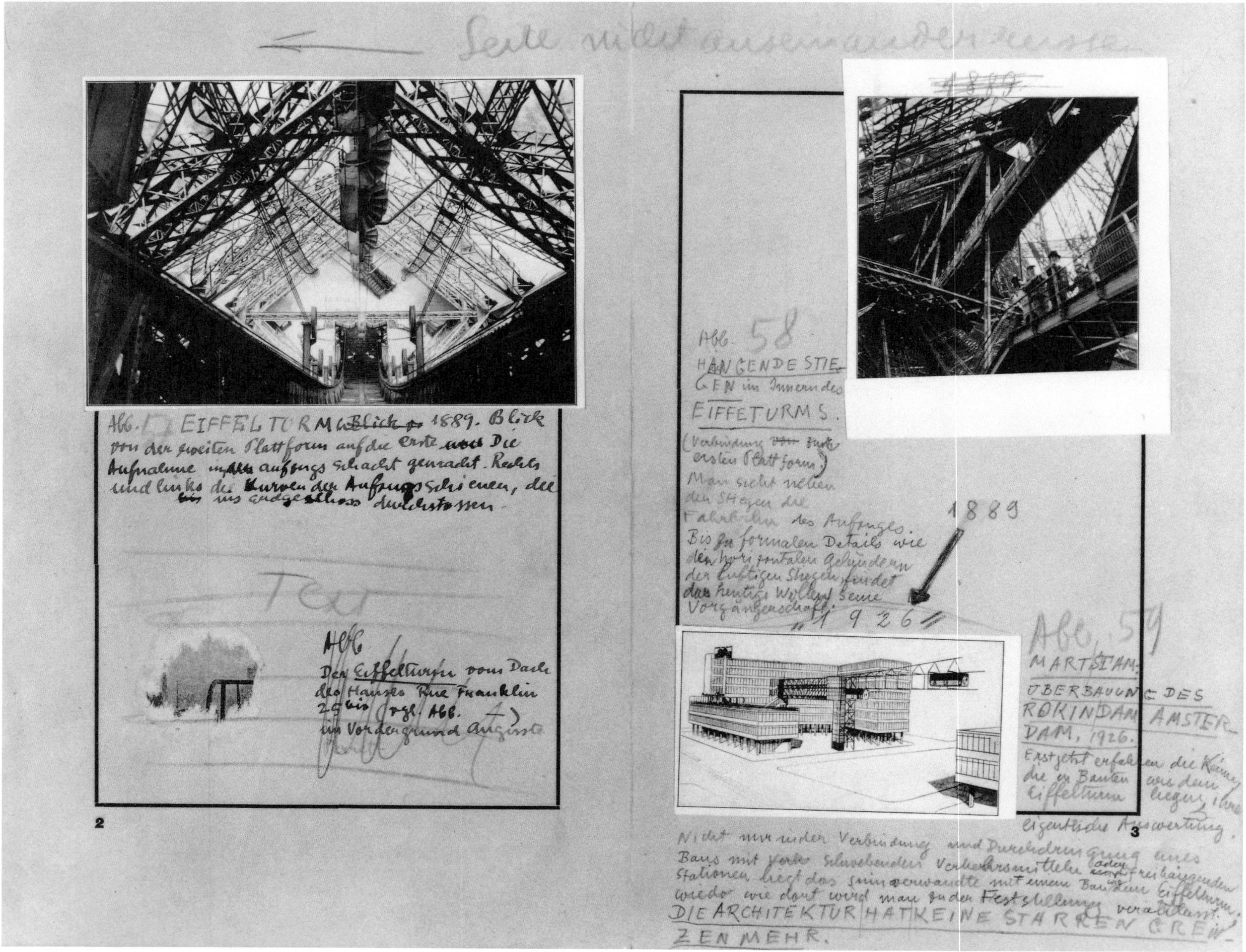

Seite nicht auseinanderreissen
Abb. EIFFELTURM 1889. Blick von der zweiten Plattform auf die erste. Die Aufnahme im Aufzugsschacht gemacht. Rechts und links die Kurven der Aufzugsschienen, die ins Endgeschoss durchstossen.
Text
Abb.
Der Eiffelturm vom Dach des Hauses Rue Franklin 25 bis (vgl. Abb. 4)
im Vordergrund [illegible]
2
Abb. 58
HÄNGENDE STIEGEN im Innern des EIFFELTURMS.
(Verbindung zur ersten Plattform)
Man sieht neben den Stiegen die Fahrbahn des Aufzugs.
Bis zu formalen Details wie den horizontalen Gehängen der luftigen Stiegen findet das heutige Wollen seine Vergangenheit.
1889
1926
Abb. 59
MART STAM:
ÜBERBAUUNG DES ROKIN DAM AMSTERDAM, 1926.
Erst jetzt erfahren die Keime die im Bauen aus dem Eiffelturm liegen, ihre eigentliche Auswertung.
3
Nicht nur in der Verbindung und Durchdringung eines Baus mit schwebenden Verkehrsmitteln oder Stationen liegt das Sinnverwandte mit einem Bau wie dem Eiffelturm. Wieder wie dort wird man zu der Feststellung veranlasst:
DIE ARCHITEKTUR HAT KEINE STARREN GRENZEN MEHR.

ABB. PONT TRANSBORDEUR, MARSEILLE

KONSTRUKTEUR: ARNODIN
Die Schwebefähre ist in der Höhe von 51 m mit der, auf dem Laufsteg gleitenden Katze verbunden. ~~Länge des Steges 240 m.~~ Die Fähre gleitet wenige Meter über dem Wasser. Länge des Steges 240 m.

Reizvolle Verbindung von festen und beweglichen Teilen.

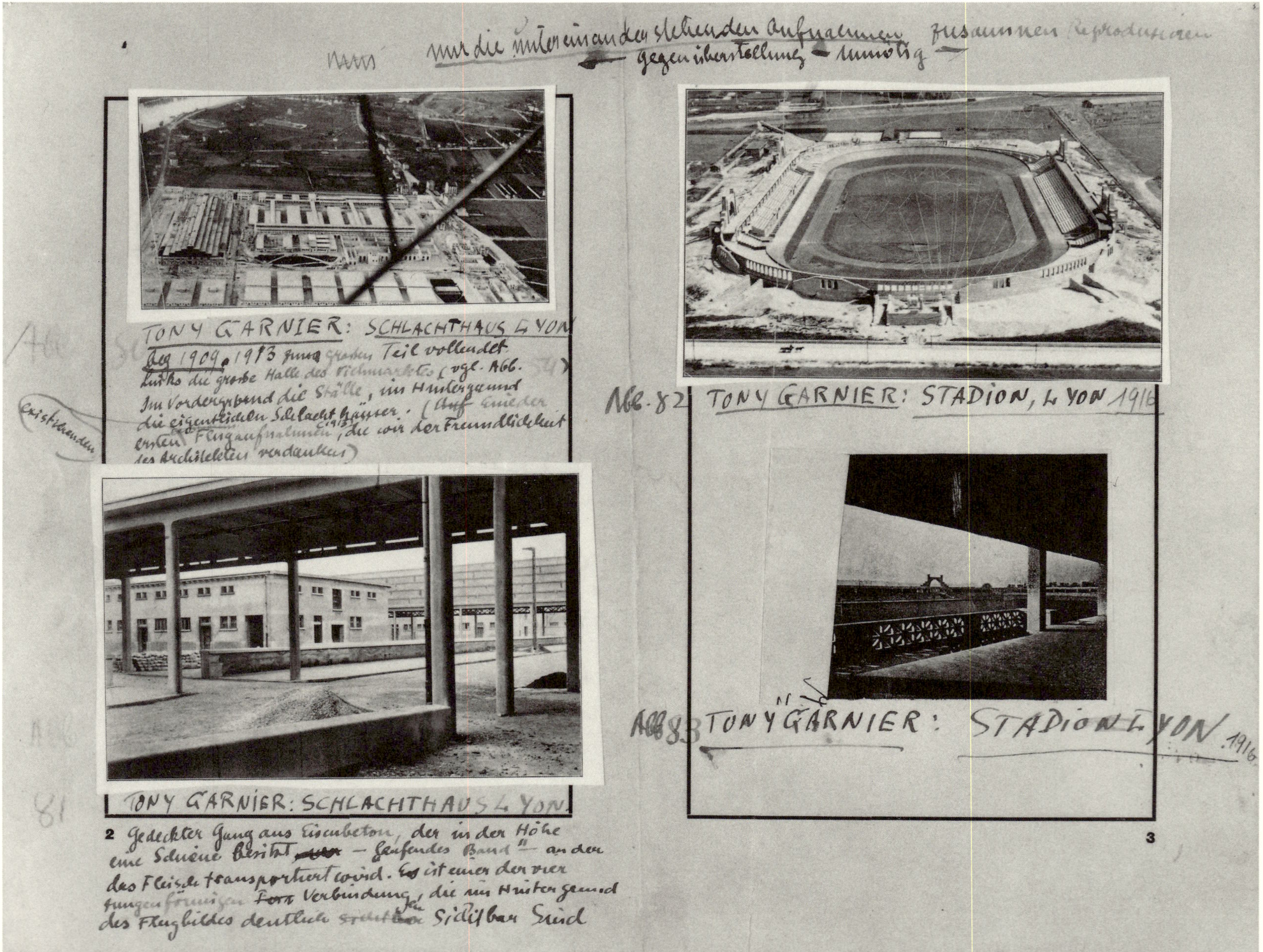

nur die untereinander stehenden Aufnahmen zusammen Reproduzieren
Gegenüberstellung unnötig
TONY GARNIER: SCHLACHTHAUS LYON
1909–1913 zum großen Teil vollendet
Links die grosse Halle des Viehmarktes (vgl. Abb.
Im Vordergrund die Ställe, im Hintergrund
die eigentlichen Schlachthäuser. (Aus einer der
ersten Flugaufnahmen, die wir der Freundlichkeit
des Architekten verdanken)
existierenden
TONY GARNIER: SCHLACHTHAUS LYON
2 Gedeckter Gang aus Eisenbeton, der in der Höhe
eine Schiene besitzt – laufendes Band – an der
das Fleisch transportiert wird. Es ist einer der vier
fingerförmigen Verbindungen, die im Hintergrund
des Flugbildes deutlich sichtbar sind
Abb. 82 TONY GARNIER: STADION, LYON 1916
Abb. 83 TONY GARNIER: STADION LYON 1916
3

Illustration Credits

Every effort has been made to locate all copyright holders for the images reproduced in this book. However, the publishers invite any copyright holders they have not been able to reach to contact them so that full acknowledgment may be given in subsequent editions.

The following sources granted permission to reproduce illustrations in this book:

Archiv S. Giedion, Institut für Geschichte und Theorie der Architektur, ETH-Hönggerberg, Zurich: figs I–VI, photograph of the cover of the German edition of *Bauen in Frankreich, Bauen in Eisen, Bauen in Eisenbeton,* figs. 1–4, 16, 20, 46, 56–58, 60–65, 67, 71, 77, 84–86, 89–90, 93–95, 97–98, 100–101, 111, 113–14, 124–25, 139, and all the illustrations in the Appendix.

Artists Rights Society (ARS), New York/SPADEM, Paris: figs. 11, 70, 72–76, 87–88, 91, 96, 99, 102–7, 109, 112, 115–22.

Bibliothèque Municipale de Lyon: figs. 80–83.

Bibliothèque Nationale de France, Paris: fig. 7.

Bibliothèque Sainte-Geneviève, Paris: figs. 10, 13.

Institut Français d'Architecture, Paris: figs. 123, 128–29.

Estate of Lucia Moholy, © Dr. F. Karsten, London: fig. 45.

Nasjonalgalleriet, Oslo: fig. 33.

Sigfried Giedion: A Biographical Sketch

Giedion was born in Prague in 1888, the son of Swiss parents. Deferring to their wishes, he took a degree in mechanical engineering, though he would ultimately abandon that field in favor of art history. His most influential teacher was Heinrich Wölfflin, one of the leading art historians of the day. Under Wölfflin, Giedion completed his doctoral dissertation on late Baroque and Romantic classicism at the University of Munich in 1922.

Rather than embark on an academic career, Giedion launched into journalism. From 1923 to 1928 he wrote prolifically for both newspapers and journals on a broad range of topics relating to the "Modern"; these included visual perception, modern art, the *De Stijl* movement in Holland, furniture design, and architecture. In 1923 he visited the Bauhaus exhibition in Weimar where he made the acquaintance of Walter Gropius. Subsequently Giedion published an article in the Swiss architectural journal *Werk* in which he depicted the exhibition not as a showcase for a faddish phenomenon but as the embodiment of an idea with unlimited potential. Strongly influenced by his social convictions, Giedion sought to promote the development of an architecture that met modern social, constructional, and aesthetic requirements. In his pursuit of this goal he became editor for modern architecture at the journal *Cicerone* in January 1927.

Giedion's encounter with Gropius, his later meeting with Le Corbusier, and his exposure to various aspects of the Modern Movement all contributed to the development of his view that modern architecture was the combination of an artistic vision and a reliance on the new materials and construction methods developed in the nineteenth century. Giedion met Le Corbusier on a visit to the *Esprit Nouveau* exhibition in Paris in 1925. The encounter so influenced him that he focused his book *Bauen in Frankreich, Bauen in Eisen, Bauen in Eisenbeton* (1928) on France, sketching the outline of a development that culminated in the work of Le Corbusier. In the process Giedion made the essential link between the historical and the critical appreciation of modern architecture.

The meeting with Le Corbusier also led to an alliance that resulted, in June 1928, in the first CIAM congress (Congrès Internationaux d'Architecture Moderne). Le Corbusier selected the architects who were invited to the congress; Giedion's task was "to keep the often confused paths of the present distinct from each other." At the congress, Giedion was made secretary general of the organization, a post he held until its dissolution in 1956. More than anyone else—Giedion was responsible for the continuity of CIAM. Through the years he managed to hold together the organization's diverse membership, mediating between a variety of ideas and approaches to architecture within an ever-changing context.

During the academic year 1938–1939 Giedion held the Charles Eliot Norton professorship at Harvard University. The lectures he gave there were subsequently published in *Space, Time, and Architecture* (1941). One of the most influential books on architecture to be written in the twentieth century, it was described by Walter Gropius as "the standard work on the development of modern architecture."

In April 1940 Giedion returned to Europe, although he longed to come back to the United States. He therefore welcomed an invitation to deliver the Trowbridge lectures at Yale University and spent 1941–1945 in the United States, returning to Zurich in December 1945. During his stay in America, he lectured and traveled widely and worked on the book *Mechanization Takes Command* (1948).

America gave Giedion the opportunity to expound his ideas on modern architecture in an academic setting, and America embraced his message. In his native country, however, he never received the academic recognition for which he yearned. He remained an outsider, even in Zurich, where he made his home from the early 1920s to his death. When he finally obtained a post at the Eidgenössische Technische Hochschule (ETH) there in 1948, it was only as a *Privatdozent* (outside lecturer). He taught at the ETH until 1958, and from the mid-1950s through the early 1960s he also regularly gave seminars at the Graduate School of Design at Harvard. In the late 1940s Giedion started working on an ambitious project trying to combine historical and archaeological knowledge with the concerns of contemporary art and architecture. This enterprise resulted in the publication of three books, *The Eternal Present: The Beginnings of Art* (1962), *The Eternal Present: The Beginnings of Architecture* (1964), and *Architecture and the Phenomena of Transition* (1971). Giedion died in Zurich in 1968.

Index

Note: Italic page numbers indicate illustrations.

Académie des Beaux-Arts, 100, 187, 199; Baudot's opposition to, 95n, 96n
Adler, Dankmar, 144n
Adler, Leo, 52
Aktivismus literary movement, 2
Alban Photographic Studio (Guévrékian), 197
Alfa-Romeo Garage (Mallet-Stevens), 190
Altes Museum (Schinkel), 120
Amsterdam: shopping street (Staal), *133*, 134; Rokin Dam (Stam), *145*
Arnodin, Ferdinand-Joseph: *pont transbordeur*, 43, *90*, *146–48*
Arts and crafts, 96n, 190
Asylum for the homeless, Paris (Le Corbusier), *183*, *184*
Auburtin, Jean-Marcel: Pleyel Hall, 194n
Autumn Salon, 1926: Beaugé design exhibited in, 198, *198*

Backer, Sir Benjamin: Firth of Forth bridge, *141*, 143
Ballot, Charles, 88n
Baltard, Victor, 97, 114; Les Halles design, *112*, 190; Saint-Augustin, church of, 114n
Banham, Reyner, 1
Banque de France: vault (Brice & Sainrapt), *202*
Barlow, William Henry, 135n
Baroque, 18, 19, 100, 120, 139, 169
Basilica of Jeanne d'Arc (Perret), 157n, 160n
Baudot, Anatole, 95, 96, 157n; Saint Jean de Montmartre (Saint-Jean-l'Evangéliste), church of (Paris), 96n, 157n, *160*
Bauhaus, Dessau (Gropius), 52, *133*
Bauhaus exhibition of 1923, 1, 225
Bazard, Amand, 97n
Beaugé, J., 194, 198; housing settlement, 198, *198*
Behne, Adolf, 52
Behrendt, Walter Curt, 41, 43
Behrens, Peter, 86
Bélanger, François-Joseph: Halle au Blé, *104*
Bellon, pont du. *See* Bridges
Benjamin, Walter, 53
Berlage, Hendrik Petrus, 86, 153, 176n
Berlin Gasworks (Schwedler), 98n
Bessemer, Sir Henry, 110
Bibliothèque Nationale, Paris (Labrouste), 14, 104n, 109
Bibliothèque Sainte-Geneviève (Labrouste), 9, 98, 103, *106*, 107, *108*
Bier, Justus, 52
Biermann, Georg, 44–45, 56, 57
Bohnstedt, Ludwig, 9–10, 20; "Über die Bedeutung des Eisens für die Baukunst," 9
Boileau, Louis-Auguste, 103, 114, 124
Boileau, Louis-Charles, 117, 132; Bon Marché Department Store, *116*, 117–18, *117*, *118*
Bon Marché Department Store (Boileau and Eiffel), *116–18*, 132
Bötticher, Carl Gottlieb Wilhelm, 5–8, 12–13, 16, 19, 24, 28, 43; *Die Tektonik der Hellenen*, 5–7, 12, 16
Bourdais, Jules-Désiré: National Opera design, 127n; Trocadéro, 127
Boyer-Gérante, R., 194; villa in Hyères, 197, *197;* Ville de Curepipe, 197
Brest, bridge near. *See* Bridges
Breuer, Marcel, 3
Brice & Sainrapt: Banque de France vault, *202;* Club athlétique de la Société générale, tribune, *202;* wine storage, *203*
Bridges: near Brest (Freyssinet), 200; over Douro (Eiffel), 98, 120n; over Firth of

Forth (Fowler and Backer), *141*, 143; Pont de Longon, *123;* Pont du Bellon, *131;* Pont du Carrousel (Polonceau), 97; *pont transbordeur*, Marseilles (Arnodin), 43, *81*, *90*, *146–48;* Saint-Pierre-du-Vauvray (Freyssinet), *200*, 201; Villeneuve-sur-Lot (Limousin & Co.), *201;* Vistula bridges (Schwedler), 98n
Broggi, Carlo, 189
Brunet, F.: Halle au Blé, *104*

Café de Unie, Rotterdam (Oud), 190
Casablanca docks (Perret), 109, 157, *159*
Casino, Saint-Malo (Perret), 154n
Chaine, Henri, 96n
Chaptal, Jean-Antoine, 94
Chartres: airplane hangars (Limousin & Co.), *201*
Chatsworth: greenhouse in (Paxton), 124n
Chicago: skyscrapers, 144; World Exhibition of 1893, 135n
Chicago school, 3
CIAM. *See* Congrès Internationaux d'Architecture Moderne
Cité Seurat (Lurçat), *196*, 197
Clark, Mathieu: Théâtre Alexandrin (Saint Petersburg), 103n
Classic architecture, 5, 6, 7, 8, 19, 24
Club athlétique de la Société générale: tribune (Brice & Sainrapt), *202*
Coignet, François, 150; exhibition building, Paris 1867, *129*
Collins, Peter: *Concrete*, 40
Comte, Auguste, 96n, 97n, 98
Congrès Internationaux d'Architecture Moderne (CIAM): Giedion's role in, 1, 4, 225–26
Considère, Armand, 151
Contamin (misspelled by Giedion as Cottancin), Victor, 135
Cook house (Le Corbusier), *107*, *179*, 180, *181*
Cremers, Paul Joseph, 52
Crystal Palace, London (Paxton), 4–5, 11, 34, 103, 124; Lucae on, 11; Meyer on, 34; plan, *122*
Cubism, 43
Curepipe, Ville de (Boyer-Gérente), 197

Daly, César, 95, 111n
Darwin, Charles: influence on Heuser, 8, 16
Davioud, Gabriel-Jean-Antoine, 95; National Opera design, 127n; Trocadéro, 127
Delaborde, Henri, 187, 188n
Department stores, 42, 47, 115–19; Bon Marché (Boileau and Eiffel), *116–18*, 117, 132; Printemps (Sédille), 117, *119;* Samaritaine (Jourdain), *119;* Wertheim (Messel), 25
De Stijl, 225
Deutscher Bund Heimatschutz, 30
Deutscher Werkbund, 25, 26, 31, 40–41
Dion, Henri de, 98, 134, 139; Galerie des Machines, Paris 1878, 134, *135*
Doesburg, Theo van, 176
Dohme, Robert: *Das englische Haus*, 21
Domino housing settlement (Le Corbusier), *168*
Dorner, Alexander, 47
Douro, bridge over. *See* Bridges
Duban, Félix: Ecole des Beaux-Arts reconstruction, 14
Dutert, Charles-Louis-Ferdinand, 135

Eck, Charles-Louis-Gustave, 103
Ecole des Beaux-Arts, 100; Duban's reconstruction, 14
Ecole Polytechnique, 94, 97, 100
Eiffel, Gustave, 98, 125, 127, 200; Bon Marché Department Store, *116–18*, 117; bridge over the Douro, 98, 120n; exhibition building, Paris 1867, *128;* exhibition building, Paris 1878, 132, *132;* viaducts over the Sioule and the Garabit, 98
Eiffel Tower, 91, *91*, 102, 120, 135, 143–44, *143–45*, 187; academy opposition, 187; Muthesius praise, 25; Jordan on, 39; Meyer on, 35; Pudor on, 27–28
Emperger, Fritz von: *Handbuch für Eisenbetonbau*, 39
Empire style, 18, 124

Esprit Nouveau exhibition (Paris, 1925), 225. *See also* Exhibitions: Paris 1925
Exhibitions, 120–40; Bauhaus 1923, 1, 225; Chicago 1893, 135n; London 1851, 4–5, 34, 97, 103, 121, *122*, 124; Paris 1798, 121, 122n; Paris 1801, 121; Paris 1855, 122–24, *122, 123;* Paris 1867, 124–27, *125–29*, 150n; Paris 1878, 127, 132, *132*, 134, *135;* Paris 1889, 18, 20, 25, 34, 109, 135–39, *136–38, 140;* Paris 1900, 150n; Paris 1925, *158*, 168, 190, *191*, 197, 225; Vienna 1873, 115n

Firth of Forth bridge (Fowler and Backer), *141*, 143
Fischer, Theodor, 41
Flachat, Eugène, 97, 111n; Les Halles design, *113*, 114
Flegenheimer, Julien, 163n
Fontaine, Pierre-François-Léonard: Galerie d'Orléans, Palais-Royal (Paris), *105*
Fowler, John: bridge over the Firth of Forth, *141*, 143
Frankfurt am Main: Central Station (Schwedler), 98n, 137n; Opera House (Lucae), 10; Palm Garden, 125n
Frankl, Paul, 77n. 184
Freyssinet, Eugène, 152, 200–201; Saint-Pierre-du-Vauvray bridge, *200*, 201
Fritsch, K. E. O., 18
Frugès, Henri, 170n

Galerie des Batailles, Versailles, 105
Galerie d'Orléans. *See* Palais-Royal
Garages: Alfa-Romeo (Mallet-Stevens), 190; boulevard Raspail (Sauvage), 194, *194;* rue Marbeuf, *204;* rue Ponthieu (Perret), 155, *156*
Gare du Nord, Paris (Hittorff), 104, 115, *115;* Polonceau truss, *110*
Garnier, Tony, 42, 84, 86, 153, 161–67, 186; *Cité industrielle*, 161–63, 190; *Grands Travaux de la ville de Lyon*, 163, 163n; Grange-Blanche Hospital (Lyons), 163, *166*, 205n. 9; Quartier des Etats-Unis (Lyons), *161*, 163; slaughterhouse complex (Lyons), *141*, 163, *164*, 198, 205n. 9; stadium (Lyons), 163, *165*, 205n. 9; suburban railroad station design, *162*, 163
Gautier, Théophile, 94
Gellhorn, Alfred, 47, 51, 52
Geneva-Cornavin railroad station (Stam and Wittwer design), 163n
Giedion-Welcker, Carola, 2
Giedion, Sigfried: *Arbeit*, 2; *Architecture and the Phenomena of Transition*, 226; *Befreites Wohnen*, 2; biographical sketch, 225–26; "Eisen, Eisenbeton, Bauen in Frankreich," *46*, 47 (lecture); *Die Entstehung des heutigen Menschen*, 2, 57; *The Eternal Present: The Beginnings of Architecture*, 226; *The Eternal Present: The Beginnings of Art*, 226; *Mechanization Takes Command*, 3, 226; photograph of, *frontispiece; Space, Time, and Architecture*, 3, 226; *Spätbarocker und romantischer Klassizismus* (dissertation), 2; *Walter Gropius*, 2; "Zur Situation der französischen Architektur" (article series), 44
— *Bauen in Frankreich, Bauen in Eisen, Bauen in Eisenbeton*, 2, 3, 4, 225; advertisement for, 49, *50;* architectural historians on, 1; design of, 45, 49; French edition, 56, 57; Giedion's flyer design for, *54, 55;* lectures on subject of, *46*, 47, *48;* response to, 51–53, 56; title choice, 45, 49, 56, 57
Gilbert, L. Cass: Tower Building (New York), 144n
Glaser, Curt, 51
Göller, Adolf, 17; *Die Entstehung der architektonischen Stilformen*, 17–18
Gothic architecture, 6, 18, 19, 100, 101, 120; and iron construction, 7, 8, 124, 152
Granet, André: Pleyel Hall, 194n
Grange-Blanche Hospital, Lyons (Garnier), 163, *166*, 205n. 9
Greek architecture. *See* Classic architecture
Groos, Karl: *Aesthetik*, 39
Gropius, Walter, 1, 3, 52, 225, 226; Bauhaus Dessau, *133*

Guévrékian, Gabriel, 194, 197; Alban Photographic Studio, 197; ferroconcrete villa, *195;* garden designs, 197; Neuilly, house in, 197; Simultané boutique, 197
Gurlitt, Cornelius, 24

Halle au Blé (Bélanger and Brunet), 104, *104*
Halles (Paris). *See* Les Halles
Haussmann, Georges-Eugène, 114
Hennebique, François, 150, 151, 154n
Herriot, Edouard, 163
Heuser, Georg, 16–17, 28
Hilberseimer, Ludwig. *See* Vischer, Julius, and Ludwig Hilberseimer
Hildebrandt, Hans, 168n
Hittorff, Jacques-Ignace, 115; "Cirque de l'Impératrice," 115n; Gare du Nord, 104, 115, *115;* "Rotonde du Panorama," 115n; Théâtre de l'Ambigu-Comique, 103
Hoffmann, Josef: Stoclet house, 190, 191
Hofmann, Albert, 20–23, 25, 26
Holabird, William, 144n
Horeau, Hector, 97, 111n, 114; Les Halles design, *113,* 114
Hyatt, Thaddeus, 150

Il Salone vaults (Padua), 124
Imperial Post Office, Berlin (Neumann), 17

Japanese art: Gurlitt on, 24
Jardin des Plantes, 103n, 124
Jatho, C. O., 51
Jeanneret, Charles-Edouard. *See* Le Corbusier
Jeanneret, Pierre, 188n
Jenny, William Le Baron, 144n
Jenny-Dürst, Hans, 44, 84
Jones, Owen, 34
Jordan, Dr. Hermann, 38–39
Jourdain, Frantz, 135n, 189n, 190; Samaritaine Department Store, *119*
Jugendstil, 119

Klopfer, Paul, 52, 53
Königliche Akademie des Bauwesens: competition, 37
Kostof, Spiro, 1
Krantz, Jean-Baptiste-Sébastien: Galerie des Machines (Paris 1867), 125, *128*

Labrouste, Henri, 25, 96n, 98, 103–9, 160, 187, 188, 188n; Bibliothèque Nationale, 14, 104n, 109; Bibliothèque Sainte-Geneviève, 9, 98, 103, *106,* 107, *108*
Lagrange, Joseph-Louis, 100
Lambot, Joseph-Louis, 150
Lang, Gustav, 23
La Roche house (Le Corbusier), 176n, *178,* 180
Lavezzari, E.: prison floor plan, *172*
League of Nations building (Le Corbusier), 157n, 162, *185, 186,* 187, 188, 194n
— competition, 189–90, 199; Broggi's project, 189; Nénot's project, 189; Perret's project, 160n
Lecointe, Jean-François-Joseph: Théâtre de l'Ambigu-Comique, 103
Le Corbusier (Charles-Edouard Jeanneret), 91n, 96, 154, 167–89, 190; Giedion and, 41, 42, 43, 52, 57, 78n. 187, 84, 225; asylum for the homeless, *183, 184;* Cook house, *107, 179,* 180, *181;* "Domino" housing settlement, *168; Kommende Baukunst,* 168n; La Roche house, 176n, *178,* 180; League of Nations building, 157n, 162, *185, 186,* 187, 188, 194n; Miestchaninoff house, *177;* Ozenfant's studio, 180; painting by, *170;* Pavillon de l'Esprit Nouveau, 168; *Peinture moderne,* 169; Pessac housing settlement, 43, 169, 170, *171–75,* 182; Stein-de Monzie house, 180, *182,* 184; Stuttgart, Weissenhof housing exhibition, 176; Swiss Jura (Schwob house), 168; *Urbanisme,* 168n; *Vers une architecture,* 41, 72n. 138, 168n, 170n, 176n, 187; villa on the sea, *178. See also* Ozenfant, Amédée, and Jeanneret, Charles-Edouard
Leibnitz, Heinrich, 7–8
Le Raincy, church at (Perret), 109, 157, *159*
Les Halles, Paris, 97, 111–14, *112–13,* 190
Lesseps, Viscount Ferdinand de, 97n

Limousin & Co., 201; Chartres hangars, *201;* Orly hangar, 152, *152;* Saint-Pierre-du-Vauvray bridge, *200,* 201; Villeneuve-sur-Lot bridge, *201*
Lindner, Werner, 30; *Die Ingenieurbauten in ihrer guten Gestaltung,* 30
Lipsius, Constantin, 13–15, 20; "Über die ästhetische Behandlung des Eisen im Hochbau," 13
London World Exhibition of 1851. *See* Exhibitions
Longon, pont de. *See* Bridges
Loos, Adolf, 153
Louis, Victor, 104n; Théâtre-Français truss, *104*
Lucae, Richard, 10–13, 20, 30, 34; Frankfurt Opera House, 10
Lucas, Felix-Benjamin: on exhibition halls, 120n
Lurçat, André, 194, 197; Cité Seurat houses, *196,* 197; Guggenbuhl house, *196,* 197; Versailles houses, 197
Lux, Joseph August, 30–33, 37; *Ingenieur-Aesthetik,* 30, 33
Lyon, Gustave, 189, 194n

Madeleine market hall (Veugny), *105*
Maillart, Robert, 151
Mallet-Stevens, Robert, 47, 190; Alfa-Romeo garage, 190; *Une cité moderne,* 190; Paris exhibition 1925, transportation pavilion, 190; Paris exhibition 1925, Bureau du tourisme, *191;* Paris exhibition 1925, concrete trees, 197; Poiret villa, *191;* rue Mallet-Stevens, 190, *192*
Manet, Edouard: *L'Exposition universelle, Paris, 126*
Marinetti, Filippo Tommaso, 30
Market halls. *See* Les Halles
Marseilles: industrial landscape, *92; pont transbordeur* and harbor, *90*
Mathon, J. B.: Pleyel Hall, 194n
Mecenseffy, E. von, 39–40; *Die künstlerische Gestaltung der Eisenbetonbauten,* 39
Menier Chocolate Factory (Saulnier), *130*
Messel, Alfred: Wertheim Department Store (Berlin), 25
Metzger, Eduard, 7
Meyer, Alfred Gotthold, 33–36; *Eisenbauten,* 33, 102n
Meyer, Peter, 41, 51
Miestchaninoff house (Le Corbusier), *177*
Mies van der Rohe, Ludwig: Stuttgart Werkbundsiedlung apartment block, *131,* 163n
Mirbeau, Octave, 96, 102
Modern Movement: Giedion and, 1, 2, 4, 225
Moholy-Nagy, László, 49, 52, 83; jacket design, *81*
Mondrian, Piet, 176
Monge, Gaspard, 100
Monier, Joseph, 150, 151
Monzie house (Le Corbusier), 180, *182,* 184
Moorish style, 15
Moreux, Jean-Charles, 194
Muthesius, Hermann, 25–26, 38; *Stilarchitektur und Baukunst,* 25

Narjoux, Félix, 114
National Opera, design for (Davioud and Bourdais), 127n
Naumann, Friedrich, 26–27
Nénot, Henri-Paul, 189
Neoplasticism, 43
Neorenaissance, 100
Neumann, Robert: Imperial Post Office (Berlin), 17
Neutra, Richard, 188n
Noailles, comte de: Guévrékian's gardens for, 197

Orly hangar (Limousin & Co.), 152, *152*
Osborn, Max, 52
Osthaus, Karl Ernst, 41
Otten, Frank, 57
Oud, Johannes Jacobus Pieter, 163n, 186; Café de Unie (Rotterdam), 190
Ozenfant, Amédée, 187; studio for (Le Corbusier), 180
Ozenfant, Amédée, and Charles-Edouard Jeanneret: *Peinture moderne,* 169

Palais-Royal: Galerie d'Orléans (Fontaine), *105*
Palm Garden (Frankfurt am Main), 125n
Panama Canal, 97n, 98
Panthéon (Paris), 94
Paris: exhibitions. *See* Exhibitions
Pavillon de l'Esprit Nouveau, Paris 1925 (Le Corbusier), 168
Paxton, Sir Joseph, 25, 98; Crystal Palace, 4–5, 11, 34, 103, *122*, 124; greenhouse (Chatsworth), 124n
Péreire, Emile, 97
Perret, Auguste and Gustave, 42, 84, 86, 109, 153, *153*, 154–61, 188, 205n. 8; basilica of Jeanne d'Arc, 157n, 160n; Casablanca docks, 109, 157, *159;* Casino, Saint-Malo, 154n; Decorative Arts Exhibition theater (Paris), *158;* League of Nations project, 160n; Le Raincy, church at, 109, 157, *159*, 160n; rue Ponthieu garage, 155, *156;* rue Franklin house, 151, 154, *154–56*, 160n; skyscraper projects, 160n; Théâtre des Champs-Elysées, 157, *158*, 160n; Versailles villa, 160n, *179*
Pessac housing settlement (Le Corbusier), 43, 169, 170, *171–75*, 182
Pevsner, Nikolaus, 69n. 108
Philipp, Salli, 63n. 57
Place de l'Europe (Julien), 144
Pleyel Hall (Auburtin, Granet, Mathon), 194n
Poiret, Paul: villa (Mallet-Stevens), *191*
Polonceau, Antoine, 97
Polonceau, Camille, 98, 110–11
Polonceau truss, 109, *110*, 111, *113*, 114
Ponts. *See* Bridges
Pont transbordeur: Marseilles (Arnodin), 43, *90, 146–48;* Giedion photo of, book jacket, *81*
Printemps, department store (Sédille), 117, *119*
Pudor, Heinrich, 27–28; *Die neue Architektur*, 27

Quartier des Etats-Unis, Lyons (Garnier), *161*, 163

Reinhardt, Max, 2
Renaissance architecture, 18, 19, 100
Reuleaux, Franz, 21
Reynaud, Léonce, 109
Rietveld, Gerrit Thomas, 180
Roche, Martin, 144n
Rokin Dam, Amsterdam (Stam), *145*
Romanesque architecture, 18
Rondelet, Jean-Baptiste, 33, 94
Root, John, 144n
Roth, Alfred, 3
Roth, Emil, 3

Sachlichkeit, 25; Lux on, 31, 32
Saint-Augustin, church of, Paris (Baltard), 114n
Saint-Jean-l'Evangéliste, Paris (Baudot), 96n, 157n, *160*
Saint Jean de Montmartre. *See* Saint-Jean-l'Evangéliste
Saint Pancras Station (London), 135n, *136*
Saint-Pierre-du-Vauvray bridge (Freyssinet), *200*, 201
Saint-Simon, Henri de, 88–89, 97, 97n, 120
Saint-Thérèse à Montmagny, church of, 157n
Saint-Vincent-de-Paul, church of (Hittorff), 115n
Samaritaine, department store (Jourdain), *119*
Sartoris, Alberto, 57
Saulnier, Jules: Menier Chocolate Factory, *130*
Sauvage, Henri, 149, 190, 192, 194; boulevard Raspail garage, 194, *194;* rue des Amiraux apartment house, 192, *193*
Scheffler, Karl, 36–37; *Moderne Baukunst*, 36
Schinkel, Karl Friedrich, 5; Altes Museum, 120
Schmarsow, August, 34
Schwedler, Johann Wilhelm, 98n; Berlin Gasworks, 98n; Frankfurt am Main railroad station, 98n, 137n; Vistula bridges, 98n
Sédille, Paul: Printemps Department Store, 117, *119*
Semper, Gottfried, 5, 8–9, 16, 18, 19, 24, 26, 35; *Der Stil in den technischen und tektonischen Künsten*, 8, 16

Simultané boutique (Guévrékian), 197
Skyscrapers, 144, 160n, 169
Sörgel, Herman, 52
Staal, Jan Frederick: Amsterdam shopping street, *133*
Stam, Mart, 107; Geneva-Cornavin railroad station, 163; Rokin Dam (Amsterdam), *145*; Werkbundsiedlung house, *178*
Stevens, Robert Livingston, 110
Stier, Hubert: on iron construction, 15, 20; "Rückblick auf die Entwicklung der deutschen Architektur in den letzten 50 Jahren" (lecture), 15
Stoclet house (Hoffmann), 190, 191
Stotz, Gustav, 51
Streiter, Richard, 7, 28–29; *Architektonische Zeitfragen*, 28
Stuttgart Werkbundsiedlung apartment block (Mies van der Rohe), *131*, 163n
Sullivan, Louis, 144n

Tettau, Wilhelm Freiherr von, 28, 33; "Ästhetik des Eisens," 28
Théâtre Alexandrin, Saint Petersburg (Clark), 103n
Théâtre de l'Ambigu-Comique (Hittorff and Lecomte), 103
Théâtre des Champs-Elysées (Perret), 157, *158*, 160n
Théâtre-Français (Louis), *104*
Thomas, Louis, 194
Tower Building, New York (Gilbert), 144n
Trocadéro (Davioud and Bourdais), 127

Van 't Hoff, Robert: Huis ter Heide, 167, *168*
Velde, Henry van de, 86, 153, 158
Vélodrome d'Hiver (Paris), *148–49*
Verband der deutschen Architekten- und Ingenieur-Vereine, 13, 15, 18
Versailles: Galerie des Batailles, 105; Lurçat houses, 197; villa (Perret), 160n, *179*
Veugny, M.-G.: Madeleine market hall, *105*
Viaducts over the Garabit and the Sioule (Eiffel), 98
Vienna World Exhibition of 1873: rotunda, 115n
Villeneuve-sur-Lot bridge (Limousin & Co.), *201*
Viollet-le-Duc, Eugène-Emmanuel, 14, 96n, 188; *Entretiens*, 14
Vischer, Julius, and Ludwig Hilberseimer: *Beton als Gestalter*, 42

Wagner, Otto, 24, 28; *Moderne Architektur*, 28
Wayß & Freytag, 150n
Welcker, Carola (Mrs. Siegfried Giedion), 2
Werkbundsiedlung, Stuttgart (Mies van der Rohe), *131*, 163n
Wertheim Department Store (Messel), 25
Wielemans, Alexander von, 40
Wiener Werkstätte Movement, 190
With, Karl, 51
Wittwer, Hans: Geneva-Cornavin railroad station, 163n
Wölfflin, Heinrich, 2, 56, 77n. 184, 225
World Exhibitions. *See* Exhibitions
Wright, Frank Lloyd, 3, 86, 153, 160, 167, 168, 180, 186

Building in France, Building in Iron, Building in Ferroconcrete
Introduction by Sokratis Georgiadis
Translation by J. Duncan Berry

Born in Thessaloniki, Greece, in 1949, Sokratis Georgiadis studied architecture at the Technische Universität Berlin. In the early 1980s he established an architectural practice in Greece but returned to Germany where he received his doctorate in the history of architecture from the University of Stuttgart in 1986. He was a guest lecturer at the universities of Zurich (1987–1988) and Bern (1991). In 1987 he joined the staff at the Institut für Geschichte und Theorie der Architektur at the Eidgenössische Technische Hochschule (ETH) in Zurich, where he served as curator of the Sigfried Giedion Archives. In this capacity he organized a number of exhibitions. From 1988 to 1994 he taught architectural history in the Department of Architecture of the ETH. He has published widely on issues of architectural history, theory, and criticism. His book *Sigfried Giedion: Eine intellektuelle Biographie* (1989) has recently been translated into English. In 1994 he became professor of architectural theory, preservation, and design at the Kunsthochschule Berlin-Weißensee. Georgiadis is currently professor of architectural and design history at the Staatliche Akademie der Bildenden Künste Stuttgart.

J. Duncan Berry focuses his studies on the architecture and theory of late historicism. He completed his doctorate at Brown University with a dissertation titled "The Legacy of Gottfried Semper: Studies in *Späthistorismus.*" His recent publications include the essay "From Historicism to Architectural Realism: On Some of Wagner's Sources," published in *Otto Wagner: Reflections on the Raiment of Modernity* (1993). He has held teaching positions in art and architectural history at Brown University, the School of Architecture at Roger Williams College, and the Rhode Island School of Design.

Designed by Bruce Mau with Chris Rowat
Type composed by Archetype, in Scala, Bodoni, and Grotesque.
Printed by Bowne of Toronto, on Cougar Opaque 80 lb., white and natural.
Bound by York Bookbinders Inc., Toronto.

TEXTS & DOCUMENTS
Series designed by Laurie Haycock Makela and Lorraine Wild

TEXTS & DOCUMENTS

A series of the Getty Center Publication Programs
Julia Bloomfield, Thomas F. Reese, Salvatore Settis, *Editors*
Kurt W. Forster, *Consultative Editor*

Building in France, Building in Iron, Building in Ferroconcrete
With the publication in 1941 of *Space, Time, and Architecture,* Sigfried Giedion captured the ideology and visage of architectural Modernism with a success that perhaps no other book in the twentieth century has matched. But the leading themes and, indeed, much of the text of that book were not progenies of the late 1930s but rather a reiteration of ideas and formulas that Giedion had rehearsed in his book of 1928, *Building in France, Building in Iron, Building in Ferroconcrete.* This book, which now appears for the first time in English, is at the same time a polemical stroke of genius and the defining moment in Giedion's life. With it, not only did he surrender in part his earlier interests in art history (nurtured under Heinrich Wölfflin), but he also positioned himself as an eloquent advocate of modern architecture. The alliance to which this book attests, together with its principles, helped to shape the direction of Modernism for the next four decades. *Building in France* was the first book to exalt Le Corbusier in an unabashed way as the artistic champion of the new movement—at the expense of a considerable body of Germanic theory and practice. It also spelled out many of the historical "myths" of Modernism such as the impoverishment of nineteenth-century architectural thinking and practice, the contrasting vigor of engineering innovations, and the notion of Modernism as technologically preordained. The very successful track of European Modernism is here given vivid form.

Otto Wagner, *Modern Architecture* (1902)
Introduction by Harry Francis Mallgrave
(Hardback, ISBN 0-226-86938-5. Paperback, ISBN 0-226-86939-3)

Heinrich Hübsch, Rudolf Wiegmann, Carl Albert Rosenthal, Johann Heinrich Wolff, and Carl Gottlieb Wilhelm Bötticher, *In What Style Should We Build? The German Debate on Architectural Style* (1828–1847)
Introduction by Wolfgang Herrmann
(Hardback, ISBN 0-89236-199-9. Paperback, ISBN 0-89236-198-0)

Nicolas Le Camus de Mézières, *The Genius of Architecture; or, The Analogy of That Art with Our Sensations* (1780)
Introduction by Robin Middleton
(Hardback, ISBN 0-89236-234-0. Paperback, ISBN 0-89236-235-9)

Claude Perrault, *Ordonnance for the Five Kinds of Columns after the Method of the Ancients* (1683)
Introduction by Alberto Pérez-Gómez
(Hardback, ISBN 0-89236-232-4. Paperback, ISBN 0-89236-233-2)

Robert Vischer, Conrad Fiedler, Heinrich Wölfflin, Adolf Göller, Adolf Hildebrand, and August Schmarsow, *Empathy, Form, and Space: Problems in German Aesthetics, 1873–1893*
Introduction by Harry Francis Mallgrave and Eleftherios Ikonomou
(Hardback, ISBN 0-89236-260-x. Paperback, ISBN 0-89236-259-6)

Friedrich Gilly: Essays on Architecture, 1796–1799
Introduction by Fritz Neumeyer
(Hardback, ISBN 0-89236-280-4. Paperback, ISBN 0-89236-281-2)

Hermann Muthesius, *Style-Architecture and Building-Art: Transformations of Architecture in the Nineteenth Century and Its Present Condition* (1902)
Introduction by Stanford Anderson
(Hardback, ISBN 0-89236-282-0. Paperback, ISBN 0-89236-283-9)

In Preparation

H. P. Berlage: Thoughts on Style, 1886–1909
Introduction by Iain Boyd Whyte

Adolf Behne, *The Modern Functional Building* (1926)
Introduction by Rosemarie Haag Bletter

LIBRARY OF CONGRESS CATALOGING-IN-PUBLICATION DATA

Giedion, S. (Sigfried), 1888–1968

[Bauen in Frankreich, bauen in Eisen, bauen in Eisenbeton. English]

Building in France, building in iron, building in ferroconcrete/
Sigfried Giedion; introduction by Sokratis Georgiadis; translation by J. Duncan Berry.

p. cm. — (Texts & documents)

Translation of: Bauen in Frankreich, bauen in Eisen, bauen in Eisenbeton.

Includes bibliographical references and index.

ISBN 0-89236-319-3 (cloth). — ISBN 0-89236-320-7 (pbk.)

1. Functionalism (Architecture) — France. 2. Building, Iron and steel — France. 3. Reinforced concrete construction — France. 4. Architecture, Modern — 20th century — France.

I. Title. II. Series.

NA1048.5.F85G413 1995

720'.944'09041 — dc20 95-10786

CIP